Josh Landis

JEFF NESBIT was the director of public affairs for two federal science agencies and a senior communications official at the White House. Now the executive director of Climate Nexus, he is a contributing writer for *The New York Times*, *Time*, *U.S. News & World Report*, *Axios*, and *Quartz*. Nesbit is the author of *Poison Tea* in addition to dozens of novels. He lives in New York.

ALSO BY JEFF NESBIT

Poison Tea

THIS IS
THE WAY THE
WORLD ENDS

HOW DROUGHTS AND DIE-OFFS,
HEAT WAVES AND HURRICANES ARE
CONVERGING ON AMERICA

Jeff Nesbit

Picador
Thomas Dunne Books
New York

picadorusa.com • instagram.com/picador
twitter.com/picadorusa • facebook.com/picadorusa

Picador® is a U.S. registered trademark and is used by Macmillan Publishing Group,
LLC, under license from Pan Books Limited.

For book club information, please visit facebook.com/picadorbookclub
or email marketing@picadorusa.com.

The Library of Congress has cataloged the Thomas Dunne Books edition as follows:

Names: Nesbit, Jeffrey Asher, author.
Title: This is the way the world ends : how droughts and die-offs, heat waves and
    hurricanes are converging on America / Jeff Nesbit.
Description: First edition. | New York : Thomas Dunne Books/St. Martin's Press, 2018. |
    Includes bibliographical references and index.
Identifiers: LCCN 2018011503 | ISBN 9781250160461 (hardcover) | ISBN
    9781250160478 (ebook)
Subjects: LCSH: Global environmental change. | Environmental policy—Moral and
    ethical aspects.
Classification: LCC GE149 .N47 2018 | DDC 363.7—dc23
LC record available at https://lccn.loc.gov/2018011503

Picador Paperback ISBN 978-1-250-23862-7

Our books may be purchased in bulk for promotional, educational, or business use.
Please contact your local bookseller or the Macmillan Corporate and Premium
Sales Department at 1-800-221-7945, extension 5442, or by email at
MacmillanSpecialMarkets@macmillan.com.

First published by Thomas Dunne Books, an imprint of St. Martin's Publishing Group

First Picador Edition: September 2019

10 9 8 7 6 5 4 3 2 1

For Micah,

who constantly reminds me of the infinite joy of learning

CONTENTS

Introduction

Harbinger

Our world is in trouble.

While it may not appear that way in America quite yet, the signs are emerging here and there if we're willing to look closely enough. Spring planting seasons start just a bit earlier and are slightly more chaotic. Extended droughts in Texas and parts of the Southwest sometimes force cattle farmers to move entire herds north for water. Tiny creatures like the American pika are running out of options in the West because they've already moved higher up on their mountains. Police cars in South Florida now have to be protected for the first time to keep salt water from rusting them out below.

Each one is a small, barely perceptible change in an America with bigger problems at the moment. More immediate questions stare us in the face when we wake each day. Can I make the rent payment at the end of the month? What if my car breaks down, and I can't get to work? What if I get really sick and can't pay my health-care bills, or I can't visit a free health clinic for advice and treatment because it's now hundreds of miles away? Will I ever be able to save any money? Should I be worried that people who aren't like me start taking jobs where I work or move into my neighborhood? Should I limit places I visit because I might be attacked? Are my kids getting the kind of education they need?

These are real, immediate concerns. They crowd in on our day, pressing on us from all sides. We don't have time to wonder or even worry about Monarch butterflies that need to keep moving north because their habitats are being disrupted. We have no sympathy for wealthy patrons of exclusive ski resorts whose favorite mountaintop playgrounds teeter on the edge of insolvency because snowpacks are no longer predictable during the winter.

It isn't our problem that serious water resource issues in California's Central Valley now cause tens of thousands of itinerant migrant workers to walk miles just to get fresh water. That's someone else's problem and doesn't affect us. If almonds become too expensive because it takes so much water to grow them, we can always buy peanuts. If alfalfa, which also fights for increasingly scarce water resources in California's Central Valley, doesn't make sense as food for cows, can't they just find something else to grow for the source of the McDonald's hamburgers we buy at the end of the food chain?

These small changes in the American landscape hardly seem worth our time right now. No one can say with absolute certainty what's causing them. Earth is made up of millions of local ecosystems, interconnected in ways that none of us can fathom. It doesn't seem quite logical that massive amounts of industrial air pollution in a country like China halfway around the world would have any real connection to our daily lives in America. Why would it? The terrible smog that envelops Beijing or Shanghai more and more is their problem—not ours. If China buys two-thirds of all the soybeans grown in the world now and may be forced to buy *all* of them within a decade, what's that to us?

The fact that hundreds of millions of people are forced to burn wood or fossil fuels to heat their homes and cook their meals, and barely manage to live from one day to the next is someone else's problem. The growing haze from coal-fired plants near big cities in dozens of developing nations that blackens the sky even during the day is unfortunate, but not something we have the bandwidth to care about. The fact that their lives are shadowed each day by a black sky seen through the haze of smog, soot, and smoke is something other people will deal with. After all, we can see blue, sunny skies in America anytime we feel like it. We've forgotten that this blue sky was a freedom we fought for in America for decades. What seems to matter is that we have it now—not what we had to do in order to win the freedom to see that blue sky when we want.

Black carbon rises into the atmosphere from millions of homes, hut stoves, forest fires, tailpipes from trucks burning diesel oil, brick kilns, and elsewhere as soot. This black carbon is nearly as responsible for trapping heat on Earth as its unseen, invisible cousin lingering overhead in the sky—carbon dioxide—that is almost entirely a by-product of humankind's efforts to find cheap energy sources from the beginning of the Industrial Revolution. Black carbon from soot and carbon dioxide from industrial activities have together created what is essentially a black sky above us, trapping heat and causing critical changes in weather patterns, ecosystems, and Earth systems.

While the black-sky effects in America are small and imperceptible right now, that isn't the case in other parts of the world. There are countries in southern Africa where extended, crippling droughts have forever altered the way families fight to live. Mothers in Madagascar are forced to boil cactus, the only plant that will grow now, to feed their children. Since cactus has virtually no nutritional value, those children regularly die of malnutrition. But it's all mothers have to give their offspring with deformed, extended bellies—that, and the occasional soup from boiled soot and ashes. Yemen is paralyzed by the world's first true civil war over access to fresh water. All but two of the country's aquifers have run dry, prompting armed conflict to protect the sources of water still there.

Farther north in Africa, across the Sahel region that is seeing its desert expand, farmers are abandoning their land in the face of questionable growing seasons and sources of water and becoming part of the flood of refugees fleeing from conflict, violence, and economic uncertainty. These refugees don't distinguish between their existential problems. They simply know they can die of environmental devastation, starvation, military conflicts, or economic collapse. They're all part of one very large, threatening landscape forcing them from their homes into refugee camps or cities with no jobs.

A billion people in India and elsewhere who still don't have access to basic things like electricity are beginning to wonder if the monsoon season that provides all of their fresh water might be in trouble. If it is, where will they get their water? Likewise, in China and parts of Asia where thousands of disappearing glaciers at the top of the Third Pole feed into nearly a dozen massive rivers that also provide fresh water for hundreds of millions of people, leaders in those countries are now fully prepared to defend their right to stop, alter, or steer those sources of water to assure that their people can live.

It isn't only people who are affected. Tens of thousands of species who can't tell us about their lives or their habitats are under severe pressure. Routine snapshots of hundreds of species in studies by biodiversity experts are starting to show "local extinctions"—when a certain kind of species can't move fast enough to adapt to its changing environment and so simply disappears for good at that place on Earth. One recent snapshot of a thousand species found that half of them were already experiencing local extinctions. Big, iconic species like giraffes and elephants are in trouble as well. They're hunted ruthlessly even as their natural habitats are threatened through environmental changes and development. They don't know why they're in trouble. But we do. Some of the creatures that every child in America has heard of could be gone from their natural homes in just ten or fifteen years. Extinction rates for all manner of species on Earth are a hundred times higher than normal.

Twenty years ago, only a very small percentage of the earth experienced extreme conditions—like a hot spell that could literally kill us if we tried to stay outside for an extended period of time. Back then, only half of 1 percent of the earth experienced such extreme events that made our place on the planet dangerous or deadly. Today, 10 percent of the Northern Hemisphere is experiencing such extreme heat events. The years of 2014, 2015, 2016, and 2017 were the hottest in the history of recorded civilization. Cities in the Middle East saw spikes of temperatures during the summer that reached almost 130 degrees Fahrenheit during the day—the sort of temperatures that make a place uninhabitable for humans if they last too long and occur too often. The central part of Australia now becomes so hot during the summer that weather forecasters had to add a new category to describe it. That region is now essentially unfit for human living during the summer.

Ecosystems in oceans around the world are under severe duress from a shift in the chemistry of the water that both depletes oxygen and makes it more acidic. Heat is killing vast swaths of living coral reefs off the coasts of nations and islands, threatening the livelihoods of fishermen who make their living from these places and countless species that also make their homes in and around these dying reefs. A big section of the Great Barrier Reef may now be lost forever.

But none of these cataclysmic changes are happening in America. We don't *see* children dying in southern Africa because extended droughts mean that their only food source is boiled cactus, a plant not nutritious enough to keep them alive. We don't see China building a three thousand–mile railroad through the Amazon to make sure they can import soybeans from Brazil and other countries in South America because they need them to feed their people. We can safely keep refugees fleeing the Sahel region in Africa from ever coming into America because it's much easier to make that issue Europe's refugee problem.

Only small changes are occurring in America—not big enough to cause anyone to even wonder why they're happening or what they might mean in the near future. Most of the Western world, specifically the United States and all of Europe, fall within the "temperate" zones of Earth and are less prone to the immediate impacts from climate change that other parts of the world are experiencing presently. Only dry regions in the American Southwest—specifically where the Colorado River provides much of the fresh water to several states—are seeing larger impacts occurring now.

Sure, for fishermen in North Carolina who have to travel hours north to fill their allotment because their catch is migrating farther and farther north, it is an inconvenience and a waste of fuel—not a catastrophe. New England lobstermen who plan to give up a way of life that has been in their families for generations because the sea is changing is a small thing, hardly worthy of the nation's attention. Farmers in the Midwest deal with more uncertainty during the planting season, but the situation isn't a crisis by any means. Bird-watchers in all fifty states know that dozens of species are at risk because their habitats are changing, but that small concern is only part of a passionate hobby. There are other bigger problems currently facing America.

But here's the thing. All of these seemingly random events are connected. The small changes happening are harbingers for something bigger. More consequential changes are coming much sooner than we know. Impacts to ecosystems great and small are occurring in every corner of the planet right now. Tens of thousands of scientists have devoted their lives to studying and chronicling these changes as specifically and carefully as they can. They might not always tell the rest of us what the harbingers truly mean at the moment, but their studies are done for the best of reasons and with the purest of motives. Those scientists are seeking the truth about what is happening to a world that is clearly in trouble. Their most sincere hope is that other people understand what they've discovered.

We are living under a black sky, in a time of uncertainty. Important changes, great and small, are happening all around us now. Those at our doorstep, or outside our window, may seem small and inconsequential to us at present. They may seem invisible to us, like the carbon dioxide that is accumulating in the black sky above us, but they will grow year by year until we've reached the limit of what our atmosphere can hold to sustain life as we know it.

We do indeed have a defined limit of what our black sky can hold in its carbon budget. Once we reach that limit, which could happen in just twenty years or so, those same tens of thousands of scientists, who have meticulously studied the chemical, biological, geologic, agricultural, and atmospheric effects we've imposed on Earth's system for decades, have collectively told us one very clear thing: When we reach the full extent of the carbon budget in a black sky above us in two decades or so, some catastrophically bad results that no one can predict might happen on the other side. The history of the earth says that breaking points have happened beyond that red line in the past. No one knows when, or what, might happen if our black sky above us reaches a point of no return. But small changes now portend such large-scale unpredictable events later.

While it's uncomfortable for scientists and experts to talk about those changes and impacts, nevertheless it's important that we try to describe them the best we can. We have to understand the limits of the black sky above us. We have no other choice, with the survival of the human race hanging in the balance. The changes and impacts of rising CO_2 greenhouse gases won't happen somewhere off in the distant future, when next generations might theoretically be able to create technology that can solve the problems. Instead, changes are happening and accelerating *right now*. Impacts, both small and large, can irrefutably be felt across our planet.

Though these changes may not affect us much personally—at least not yet—it's critical that we care, hear, and understand the real stories of climate change across the globe as harbingers of a much bigger reality. The poignant stories in this book are all tales of our shared humanity—what we're truly facing on a planet we've conquered at a very costly expense. Admittedly, some of the stories are disturbing, even terrifying . . . even more so because we humans, as a species, have created nearly all of the problems and crises ourselves. However, the good news is that we also have the capacity to solve those problems and crises before they reach a point of no return, if we choose to act now.

This Is the Way the World Ends is meant to change the current worn-out, redundant, polarizing climate change arguments to a united climate conversation that's relevant to every human and focuses on a clear blueprint of real, workable solutions we can tackle together. That blueprint begins with understanding the real-life stories and scientific facts of what is happening right now across our globe. Because the subject matter is so large, we'll explore it in stages.

- "The Truth" explains what's happening in our world now—not in some remote, distant future we all have difficulty truly caring about.
- "The Ecosystems" describes how the changes scientists and others are

discovering already create large "regime shifts" in our oceans, the atmosphere, and at the ends of the earth.

- "The Impacts" uncovers how devastating these changes are to species, cities, farming, and other critical aspects of life as we know it.
- "The Geopolitics" shows what these changes mean in the real world—how they're unfolding across every continent right now and why so many countries are on the edge of despair.
- "The Blueprint" offers an inspired, realistic way out of the box canyon we now find ourselves in.
- "The Future" reveals why the human species currently faces its most important test and what we must do to succeed this time around.

Humans have passed every other challenge—from the creation of civilization in the shadow of the last Ice Age to the Industrial Revolution. Understanding the ever-growing black sky as a harbinger for future cataclysmic events—which are closer than we might think—is our next great test as a species. Our challenge today may indeed be greater than those others, because we cannot afford to fail the test, or life as we know it will end.

This Is the Way the World Ends explores the harbingers happening in our own backyard, the bigger changes occurring around the globe, and the ramifications of climate change to all of us. Once we truly know what we're seeing with our own eyes in the present, we can finally set to rest any questions about what the harbingers mean for our own immediate future.

We can then tell our leaders to do something while there is still time and use our voting power to insist that they follow through. We can make our cities brighter and cleaner in ways that bring balance and common sense back to urban landscapes. After thousands of years of trying to dominate and conquer the earth, we can end that one-sided perspective and learn new ways to coexist with our environment in a healthy way. And we can use our collective ability as consumers to make certain that companies do what's right and what's necessary to transform our planet in order to make it through a rapidly closing gate.

We really don't need to save the planet. The planet will survive. However, we, as a human species, may not unless we tackle the problems we've created on an interconnected Earth and alter the ways we interact with our global home.

The clear solutions are in front of us. All we have to do is take action. We can't afford not to.

Climate change is no longer some far-off problem;
it is happening here, it is happening now.

—Barack Obama

PART 1

The Truth

These harbingers reveal what's happening in our world right now—
not in some remote, distant future.

Irrefutable evidence that we're in trouble from Earth's changing climate surrounds us. As the burning of fossil fuels accelerates the greenhouse gas effect—trapping heat on our planet at an ever-increasing rate—we are rapidly reaching a point of no return, scientists warn.

"The Truth" explores the dramatic changes to the earth's ecosystem through rising greenhouse gas levels and what the ominous rise really means to life as we know it. The effects include more frequent, longer-lasting heat waves; rapidly melting glacial ice and resulting rising sea levels; precipitation changes that cause devastating floods and prolonged droughts; the growing extinction and migration of local species that can't adapt to the resulting environmental changes; and the acidification of oceans. Such changes already threaten the livelihood and lives of tens of millions of people in distant lands and are insidiously creeping into our own backyard.

With a shrinking Third Pole and a sixth mass extinction under way, the planet is already changing dramatically *right now*. What happens when we reach the critical CO_2 threshold of 450 parts per million (ppm)—which we could hit in just fifteen years or so if we remain on "business as usual" paths? And is there a way to rewind the clock?

In the face of such overwhelming questions, one point is clear: By any estimation, CO_2 levels at 400 parts per million are infinitely safer than 450 parts per million. Acknowledging the dangers, understanding the stories of what's happening right now, and deciding to take those threats seriously are the first steps in strategizing workable solutions.

1

Einstein's Warning

Most people have heard some version of the "Einstein letter" story—about the famous theoretical physicist Albert Einstein warning FDR that the Nazis were about to develop a nuclear bomb, which so alarmed the president that he immediately started the Manhattan Project. From there, scientists created the atomic bombs that America dropped on two of Japan's largest cities in order to end the war. That's the story people have heard, in some fashion. It's mostly wrong.

It's true that Einstein was convinced by his peers, such as Hungarian-born physicist Leo Szilard, that very smart physicists in Nazi Germany were close to the then theoretical ability to split the atom and harness its energy. Einstein and Szilard desperately wanted to get President Roosevelt's attention—to warn him of what might happen in the not-too-distant future should Nazi scientists succeed in that effort and create a weapon based on what was (at the time) a possible future built from math equations and theoretical physics.

The myth is that Einstein wrote his letter to President Roosevelt, warning him of the grave risk; that it was given to FDR soon after he'd written it; and that Roosevelt and his political and military advisors were so alarmed by it that they immediately charged the nation's top scientists with developing an atomic weapon in a race to beat the Nazis. The Einstein letter triggered high-level action from the White House, swiftly leading to the Manhattan Project and the birth of the atomic age, the story goes.

The truth is that Einstein actually wrote three such letters in 1939 and 1940—each more insistent than the previous one—laying out the potential risks should Nazi scientists succeed in acquiring an atomic weapon before the United States did. What he and Szilard described was theoretical research from German and American physicists published in several prominent journals in early 1939 that explained the potential for nuclear fission and how to exploit it to create nuclear power. Szilard concluded from the research that a nuclear weapon was possible, and he convinced Einstein.

But Einstein was mostly an irritant to FDR and his senior White House staff with his warnings. FDR had a real war to consider—not some theoretical future threat. FDR's political advisors did not believe the Nazis were close to splitting the atom and harnessing its power, and his military advisors similarly dismissed any potential.

What's more, his first letter—the one known as the famous "Einstein letter"[1]—was delayed for nearly three months for political and military reasons. It was drafted by Szilard and Einstein in late July 1939 and then dated August 2. Their plan was to give it to FDR through an intermediary they believed had FDR's ear. But Germany invaded Poland before the letter could be delivered. It wasn't actually delivered to FDR by that intermediary, Alexander Sachs, until mid-October of that year.

"In the course of the last four months it has been made probable—through the work of Joliot in France as well as Fermi and Szilard in America—that it may be possible to set up a nuclear chain reaction in a large mass of uranium, by which vast amounts of power and large quantities of new radium-like elements would be generated. Now it appears almost certain that this could be achieved in the immediate future," Einstein wrote.

"This new phenomenon would also lead to the construction of bombs, and it is conceivable—though much less certain—that extremely powerful bombs of this type may thus be constructed," he added. "A single bomb of this type, carried by boat and exploded in a port, might very well destroy the whole port together with some of the surrounding territory. However, such bombs might very well prove too heavy for transportation by air."

The immediate reaction from FDR and his senior White House staff wasn't an alarming one. It was a bureaucratic one. He sent a polite note back to Einstein—thanking him for the warning—and set up a committee headed by the director of a very small science agency (the Bureau of Standards) and two lower-level military aides to study it. FDR's response was a time-honored brush-off, typical of Washington then and now.

One of those aides—army lieutenant colonel Keith Adamson—was an ordnance specialist who had attended the meeting at the White House when Sachs delivered the first Einstein letter. When the small committee met, Adamson was one of two military advisors present. He was skeptical of the notion that an atomic weapon could be developed but signed off on a small $6,000 grant to Szilard and Enrico Fermi that allowed them to purchase uranium for an experiment at a lab that they had jointly created to test the theory.[2]

From this very humble and bureaucratic beginning, and the small grant to purchase uranium, Fermi was eventually able to prove his theory. Fermi created the first "atomic pile"—and the notion of splitting and harnessing the atom was transformed from theory to reality.

But the truth is that the government committee FDR set up after Einstein's first letter didn't lead to a vigorous effort to pursue and develop an atomic weapon. Einstein actually felt compelled to write a second letter in March of 1940 and then a third letter a month later. The truth is that a president consumed with war had no time, or need, for scientists and their apocalyptical warnings. Einstein pleaded with FDR to heed the science.

"Last year, when I realized that results of national importance might arise out of research on uranium, I thought it my duty to inform the administration of this possibility," Einstein wrote in his second letter.[3] "Since the outbreak of the war, interest in uranium has intensified in Germany. I have now learned that research there is carried out in great secrecy and that it has been extended to another of the Kaiser Wilhelm Institutes, the Institute of Physics." Einstein then warned the White House that his colleague, Szilard, was about to publish new research describing "in detail a method of setting up a chain reaction in uranium."

But the White House, and the committee FDR had set up, still didn't act on the science. So Einstein wrote a third letter in April 1940. "I am convinced as to the wisdom and the urgency of creating the conditions under which that and related work can be carried out with greater speed and on a larger scale than hitherto," Einstein wrote to FDR. He was so concerned at the slow pace and general lack of interest in developing an atomic weapon before the Nazis did that he proposed using money from "private sources" to accelerate efforts. An effort partially funded by such private sources "could be carried out much faster than through a loose cooperation of university laboratories and government departments."

FDR and his advisors never truly acted on this or any of the letters Einstein sent, where he laid out the emerging science and its implications. The committee set up by his first letter was dead-ended and superseded by two other government committees and offices in 1940 and 1941. Only when it became apparent that Winston Churchill and the British were quite serious about the pursuit of an atomic weapon[4,5] did FDR authorize full-scale development in January 1942—a full two years after the first theoretical physics research Einstein had brought to their attention was published in the peer-reviewed literature. Nuclear fission research was then taken over by a *fourth* government committee, and the U.S. Army Corps of Engineers' Manhattan Project began.[6]

While much has been made of the Einstein letter and the famous physicist's later regret at having started the process that led to the creation of the atomic bomb, the truth is that physicists in Germany, Britain, and the United States were all competing with each other and publishing in the scientific literature. It was only a matter of time before someone solved the nuclear fission equation.

The great irony—and an important point—is that Einstein's three letters show that politicians and leaders generally ignore scientists and research until something else compels them to act. In this case, FDR and the White House staff didn't really jump into full-scale development and the Manhattan Project until it became apparent that Churchill and Britain were already committed. The science, in and of itself, was not sufficient enough to force FDR's hand. Other events did so.

In an even further irony, Einstein was never allowed to work on the Manhattan Project. The army denied him the work clearance he needed to collaborate with some of his colleagues on the highly classified project. His pacifist leanings made him a security risk, the army concluded.[7] Einstein actually wrote a fourth letter to warn FDR of the risks of using atomic weapons without adequate oversight, but it didn't reach FDR in time. President Roosevelt died before Einstein's fourth letter could reach him.[8]

That's the sad, but true, story of the famous Einstein letter. Scientists are routinely ignored by political and business leaders in the United States and around the world—no matter how right and prescient they might be about any given subject.

Collectively, right now, thousands of modern-day Einsteins are yelling as loudly as they possibly can that we're in trouble on our planet from Earth's changing climate. The signs of that trouble range from massive species extinction and ocean system collapses to water scarcity and food insecurity that now threaten the lives of tens of millions of people in distant lands. But, like Einstein, they've become irritants to our collective political psyche. We'd like them to go away. Unlike FDR, we can't throw a measly $6,000 at Enrico Fermi and tell him to prove the theory of catastrophic impacts on our habitable planet before we decide whether to take the threats seriously or not.

It's an old story to the global media. It's a tired, worn-out story for the public. It's a deeply divided subject for politicians. It's a largely irrelevant story for business leaders—at least, for now—who have to deal with quarterly earnings reports for their shareholders. And it's a maddening story for scientists. Somehow, in some yet undiscovered fashion, we need to find our way out of this box canyon before it's too late.

There have been thousands upon thousands of peer-reviewed research papers written about the indisputable science of climate change. The evidence is overwhelming and irrefutable. Levels of greenhouse gases in the atmosphere are rising. Temperatures are going up. Springs are arriving earlier. Ice sheets are melting. Sea levels are rising. The patterns of rainfall and drought are changing. Heat waves are getting worse, as is extreme precipitation. The oceans are acidifying. On every continent and in every ocean, animals and plants are moving toward the poles.

The science linking human activities to climate change is almost directly analogous to the science linking smoking to lung and cardiovascular diseases. For a long time, the tobacco industry paid scientists to question the science. It took years for scientists, journalists, and public health officials to successfully counter tobacco industry claims that smoking was not necessarily responsible for lung cancer. Eventually, physicians, cardiovascular scientists, public health experts, and others all came to a consensus and agreed that smoking causes cancer. And this consensus among the health community has convinced most Americans that the health risks from smoking are real.[9,10]

But, believe it or not, we still don't *completely understand* how or why smoking leads to lung cancer. What we do know, however, is enough to conclude that smoking is, in fact, the reason that so many people die of small-cell lung cancer every year, even if we cannot show with 100 percent accuracy the physiological mechanism that leads from smoking to the advent of the cancer process in someone's lungs.[11]

A similar consensus now exists among climate scientists . . . a consensus that maintains climate change is happening, and human activity is the principal cause. Thousands of scientists said precisely that in the latest report of the UN's Intergovernmental Panel on Climate Change convened by the UN that has evolved in the same manner as the science of what we know has evolved. It is, for scientists, now beyond dispute—it is a "settled fact."

We now see with clear, peer-reviewed scientific certainty that seventeen hundred species are moving inexorably toward both poles to escape planetary warming and that half of the known species on Earth are experiencing local extinctions right now; that CO_2-driven ocean acidification is already destroying coral reefs and that big iconic reefs like Australia's Great Barrier

Reef are almost gone; that we've lost more than one thousand cubic miles of ice on the planet, which is raising sea levels, according to NASA satellites; that the number of arid areas on the planet has doubled in fifty years; that large wildfires connected to climate forces in the American West have increased sevenfold in a generation; that the number of floods (connected to extra precipitation generated in a climate driven by warming) has tripled in fifteen years; that record-high temperatures in the U.S. are now twice that of record lows; that the number of climate-driven natural catastrophes worldwide has doubled in the past thirty years; and that the level of actual ice volume in the Arctic (not just the surface area) has shrunk every single year for the past twenty-five years and is now close to disappearing during summer months.

This last fact alone—that the sea ice volume in the Arctic is now a quarter of what it was just twenty years ago and is presenting us with an ice-free Arctic Ocean eighty years earlier than the Nobel Prize–winning IPCC report predicted in 2007—should be more than enough to tell us what we need to know about climate science. The Arctic sea ice situation is as clear a science canary in the proverbial coal mine as we'll find . . . especially when we consider that what happens in the Arctic has profound implications for the rest of Earth.

But there are also other climate impact canaries on a grand scale on Earth now. The ten hottest years in the history of human civilization have occurred in the past twenty years. The years of 2014, 2015, 2016, and 2017 were the hottest years ever in human history. The rate of increase in carbon dioxide levels into the lower atmosphere have doubled every decade, beginning at the dawn of the Industrial Revolution. And, while few of us seem to care about sea level rise (because its worst effects will appear after we're all dead and gone), it is now a problem everywhere. For instance, most people don't appreciate wetlands—those freshwater spots along the coastland that we rarely visit because we don't build much on them. Between 1970 and 2008, natural coastal wetlands declined by nearly 50 percent, squeezed by development and sea level rise and saltwater intrusion.[12]

The world's changing climate story took years to understand and truly appreciate. We are only just now beginning to grasp the immediate consequences of our actions in nearly every corner of the planet.

It began with a simple question: Why does the earth breathe, and what does that mean?

A scientist, Charles Keeling, discovered this important fact about our planet in the 1950s.[13] When he very meticulously began to measure carbon dioxide levels near the top of a volcano in Hawaii, he discovered that the levels rise and fall throughout the year as the plants on Earth take in CO_2 and then shed that CO_2 as the leaves fall and decay.

But, while measuring that rise and fall, Charles Keeling also discovered a disturbing trend. CO_2 levels, averaged across the planet on an annual basis, rose a little more each year. They were at 310 parts per million when he began to measure them in the 1950s. A half century later, they'd risen to more than 400 parts per million.

The rise in carbon dioxide in Earth's atmosphere is relentless, indisputable, unmistakable, and incontrovertible. It is the necessary ingredient that allows Earth's troposphere to trap heat on the surface of the planet and in our oceans. As CO_2 rises in the lower atmosphere, heat rises on Earth. We can argue with anything we'd like when it comes to planetary warming, but one thing we can't argue with is Charles Keeling's curve.

And the Keeling curve—even to a casual, cynical, or skeptical observer—shows that we're likely to reach 450 parts million in the next ten to fifteen years unless something changes the equation on our planet.[14]

But even as Charles Keeling, a registered Republican, meticulously chronicled CO_2 levels, others took up the search for what it meant. Keeling's widow told Justin Gillis of *The New York Times* that her husband didn't believe global warming was all that political. It was just an inexorable fact—one he'd studied his entire career.

Scientists began to study past levels of carbon dioxide in air bubbles trapped in the ice in Antarctica. They found that CO_2 levels have naturally risen and fallen between 200 and 300 parts per million in the past four hundred thousand years. They also studied past temperatures from hydrogen isotopes in the same ice in Antarctica, going back four hundred thousand years. And they discovered a pattern. As carbon dioxide levels rose, so did the planet's temperature. And as CO_2 fell, the temperature did, too.[15]

But what the earth has never experienced before is the trend that Charles Keeling found. We don't know what happens in modern history when the atmosphere reaches 450 parts per million of carbon dioxide. As we continue to burn fossil fuels at an unprecedented pace, accelerating the greenhouse gas effect, we are approaching a point of no return. What happens when we get to 450 parts per million, with greenhouse gases like CO_2 trapping heat on the planet at an ever-increasing rate?

Scientists would prefer not to see that. To keep the planet safe for human inhabitation, they'd like to find a way to rewind the clock and move the CO_2 needle back to 350 parts per million. It's infinitely safer at 350 than 450. Whether that's politically or economically possible is a different question entirely . . . and leaves scientists with an interesting hypothesis that may be tested in real time. What happens when greenhouse gases in a planet's atmosphere move beyond a critical threshold?

In the very extreme, take a look at the surface of Venus to see what happens when a runaway greenhouse effect takes over. Venus is about the same size as Earth, though closer to the sun. It has an atmosphere, but it's vastly different from Earth's. About 97 percent of Venus's atmosphere is made up of carbon dioxide, compared to less than 1 percent CO_2 on Earth.

Once, there was probably water on Venus and its surface temperature was about twice that of Earth (120 degrees Fahrenheit or so). But as CO_2 took over Venus's atmosphere, temperatures began to rise. The water disappeared. Today, temperatures on Venus can reach 900 degrees Fahrenheit. That's the greenhouse effect, in extreme. On Venus, it increased the temperature to a point where the atmosphere would crush a human being and melt lead.[16]

But that's the extreme. Here on Earth—if we exceed 450 parts per million—we could move past a point at which everything accelerates beyond our control. Earth's system is a complex web. We've never pushed it to this point before in modern human history, and hazarding a guess about what might happen when CO_2 levels are 50 percent higher than they've ever been before in the recent history of the planet is just that—a guess.

If Earth's climate changes sooner than we expect, it could bring about substantial sea level increases, considerable ocean acidification, rapid warming of ice at both poles, an unprecedented change in freshwater patterns in the oceans, and eventual shifts in circulation patterns in both the atmosphere and the oceans. In other words, our basic Earth geosystem could change in very dramatic ways.

Beyond 450 parts per million, coral reefs could start dissolving. Average temperatures could rise dramatically. Half the species on our planet could disappear. Dust bowls could become commonplace. Drought and monsoons could become the norm in parts of the world. And, unfortunately, it would take the human species a thousand years—or longer—to reverse the damage.

We know what happens when greenhouse gases in the atmosphere increase. We can measure it, study it, and map it. We know what a greenhouse effect is. The question is whether we can keep that from happening on Earth—before we've gone beyond 450 parts per million and a possible point of no return.

Unfortunately for us, and the planet, the rate of increases in the amount of carbon dioxide we're putting into the atmosphere is now more than double the rate of what it was only a generation ago, according to meticulous monthly records kept at the National Oceanic and Atmospheric Administration (NOAA) and at the Mauna Loa Observatory in Hawaii, where Keeling first began his careful work studying CO_2 levels.[17] And it shows no sign of stopping.

This doesn't bode well for the planet. In fact, atmospheric CO_2 is accelerating. Increases are essentially doubling from decade to decade. Climate supercomputer models all assume the world's leaders will slow atmospheric CO_2. The opposite is happening, which means that we're now conducting a very large science experiment with our planet's atmosphere.

In a recent decade, the average, annual rate of increase in atmospheric CO_2 was 2.07 ppm. That's more than double the increase in the 1960s, when it was less than 1 ppm a year. Here are the hard numbers: CO_2 increases averaged 2.07 ppm a year from 2003 to 2012; 1.67 ppm from 1993 to 2002; 1.52 ppm from 1983 to 1992; 1.37 ppm from 1973 to 1982; and 0.90 ppm from 1963 to 1972. The current trend is the same. CO_2 levels rose 5.5 ppm in just two years, from March 2015 to March 2017 (an annual average of 2.25 ppm).

Regardless of our beliefs or skepticism, carbon dioxide levels are rising. Temperatures will rise along with them, at some rate. Lowering fossil fuel consumption and reducing greenhouse gas levels can't do harm to the planet and all of us who live on it. Allowing CO_2 and other greenhouse gases to reach critical levels in our atmosphere, however, could mean the end of life as we know it on our planet at some point.

Our best science says that 450 parts per million of carbon dioxide in Earth's atmosphere may very well be a critical threshold that we shouldn't cross. It is the point at which we reach the end of the safe carbon budget on Earth. We don't truly know what happens on the other side. As Charles Keeling might have put it, the steady march of carbon dioxide levels in our atmosphere today is an unmistakable, incontrovertible, irrefutable truth. Science has clearly shown it.

But what we choose to do about that truth is another question entirely.

2

Species on the Move

When Terry Root—a soft-spoken biodiversity expert and now a senior fellow emerita at Stanford University's Woods Institute for the Environment—decided to study whether birds in North America were flying north earlier or shifting their ranges in response to Earth's changing climate system three decades ago, she wanted to know if birds were changing their patterns on a big, continent-wide scale. At the time, others were evaluating much smaller changes in local areas, including biodiversity research of environmental factors such as predator-prey relationships. But Root focused on temperature—on a continental scale. She wanted to learn whether birds and their ranges across North America were affected by climate changes and temperature fluctuations.

What she found was that birds only migrate as far north as their metabolism will allow them to compensate for heat lost overnight. Her research firmly established that bird species *were* moving their ranges over time. They were coming north earlier in the spring and leaving later in the fall. They were moving their ranges up in elevation and latitude, toward the poles. Root's groundbreaking migratory bird study was the first definitive proof that birds and other species were responding to the changing climate and that humans were responsible for the changes.[1]

At the time, no one had yet asked the question now at the very center of hundreds of biodiversity research projects in every corner of the planet: How will Earth's changing climate system affect us and every species that lives on the planet with us? Root's research told a simple, elegant story that, yes, species *are* migrating away from a growing threat to their local habitats. Her study led the way to thousands of other biodiversity studies chronicling species great and small being forced to move from their homes toward the poles to escape a warming planet.

Root was a lead author of the first biodiversity reports issued by the global climate change body of experts (the Intergovernmental Panel for Climate Change) that told the world why species on the planet were starting to move north or south toward the poles to escape rising temperatures. She knows more about biodiversity and the impacts of Earth's changing climate system on species than just about anyone, ever.

In 2003, Root and other leading authors decided to look at every piece of research done to that point on species migration that had been conducted since her first migratory bird study. What they found was that, yes, a profound shift was under way.

"Over the past 100 years, the global average temperature has increased by approximately 0.6 degrees C[elsius] and is projected to continue to rise at a rapid rate," Root and other authors wrote to open that seminal 2003 study.

Although species have responded to climatic changes throughout their evolutionary history, a primary concern for wild species and their ecosystems is this rapid rate of change.

We gathered information on species and global warming from 143 studies for our meta-analyses. These analyses reveal a consistent temperature-related shift, or 'fingerprint', in species ranging from molluscs to mammals and from grasses to trees.

Indeed, more than 80% of the species that show changes are shifting in the direction expected on the basis of known physiological constraints of species. Consequently, the balance of evidence from these studies strongly suggests that a significant impact of global warming is already discernible in animal and plant populations.[2]

It's hard to ignore the unmistakable conclusion of this study by Root and others. By 2003, four-fifths of the species moving north or south toward the poles were doing so because of "physical constraints" on their habitats. The researchers had found the clear "fingerprints" of impacts on species and plants everywhere. It was now "discernible in animal and plant populations." Species and plants all across the planet, in every valley and mountain on Earth, were doing their best to adapt to a rapid change that they didn't understand and couldn't control—but which threatened their existence.

I interviewed Root several years after this important 2003 biodiversity study and long after her first, pivotal migratory bird research. It was part of a science video series with leading experts on what they knew definitively about Earth's changing system. I'd commissioned the series at the National Science Foundation, where I worked. It was released as the first side event of the 2009 global climate summit in Copenhagen.[3]

We set up the camera in a classroom on Stanford's campus. There were no crowds. This wasn't live television. I simply wanted to talk to her about her research over the years that had been so critical to what we know about how species are now reacting to the changes on Earth. I asked her about the migratory bird study that had started it all so many years ago. I asked her about the bigger reports over the years—the first biodiversity report for the IPCC that had established the basis for what the world knew, and about that 2003 meta-analysis that had looked at dozens of biodiversity studies on hundreds of species.[4]

She struggled to answer the questions. She would start down one path, change to another, and then stop. We tried several different approaches. After the sixth time, she broke down on camera. I turned the camera off so that we could just talk. Clearly, for whatever reason, this act of telling simple stories about what her research had found over decades of study was deeply troubling to her—to the point that she was incapable of explaining what she knew and what researchers in her field had found.

I asked her what was wrong.

"I've failed. I've tried. But I can't make people understand," she told me off camera, the immense sadness of her self-perceived failure palpable. "I don't know how to make people understand what's happening, what it means."

We talked for some time before returning to the interview. She had spent her entire professional life unraveling a complicated, profoundly disturbing science question: Were species on Earth threatened by rapid changes on land, in the sky, and in the oceans, and were they reacting or adapting to those changes? They'd found definitive answers.

But toward the end of her career, she'd come to believe that no one beyond her small world of scientific peers truly understood what it all meant. She believed—genuinely, truly believed—that she had failed in her most important task, which was to help people understand that it was real, that it was happening, and that every species on Earth was now being affected by the rapid changes. That was what was so deeply troubling to her. She wanted people to understand what she and all of her colleagues knew—and she believed that she had utterly failed.

She didn't mind that people questioned her work or that they even questioned her motivations and intentions. She was used to the brutal world of peer-reviewed science at the highest levels, where all of the smartest people in the room constantly challenged every underlying assumption, every basis of fact, and what they could definitively say about what researchers in the field were seeing and observing with their own eyes. That's the science process, which questions and questions and questions until it arrives at definitive answers.

But what bothered her so profoundly was that she and every other biodiversity expert in the world had found those answers. Species great and small—from the simple monarch butterfly to big, iconic species in Africa—were clearly in trouble. Many were facing certain extinction because they were unable to adapt fast enough to rapid changes occurring on Earth. She knew that. However, she felt utterly incapable of explaining that complex set of answers to even more complicated questions to a public that seemed unwilling, or incapable, of spending any time understanding what it all meant.

She was struggling—as all scientists do—to shrink a lifetime of work into a brief statement that conveys instant meaning to a mass audience. I told her what I've told hundreds of scientists over the years. "Don't worry. That isn't your job. You do the work. You find the answers. Take as much time as you need to explain it—why the answers that you've found are real; that others can follow what you've done and repeat the work if necessary; and that your research findings will stand the test of time. People will come to understand your work and what it really and truly means. If it takes ten thousand words to explain it properly, then take ten thousand words. It isn't your job to explain the meaning of your work in 140 characters or a thirty-second sound bite. Science always wins in the end. It just takes time."

What Terry Root and other biodiversity experts like her have found in the past thirty years is an unfolding science story that should give every single human being on Earth pause. Nearly half of all species on Earth are now experiencing "local extinctions"—which means that they can no longer live in their local habitat and are dying out for good in that spot on Earth. Others of the species might be able to adapt in other places, and the species as a whole might be able to survive, but they have died for good where they are locally.

John Wiens, a biology professor at the University of Arizona, looked at nearly 1,000 species in various parts of the world and found that "climate-related local extinctions have already occurred in hundreds of species, including 47% of the 976 species surveyed," he reported in *PLOS Biology* in December 2016.[5]

Wiens said that the local extinctions were "broadly similar across climatic zones, clades, and habitats, but was significantly higher in tropical species than in temperate species (55% versus 39%), in animals than in plants (50% versus 39%), and in freshwater habitats relative to terrestrial and marine habitats (74% versus 46% versus 51%)."

What these startling results show is that "local extinctions related to climate change are already widespread" and that "these extinctions will presumably become much more prevalent as global warming increases further by roughly 2-fold to 5-fold over the coming decades."

What Wiens's study shows is that, as the planet warms, species around the world are racing against time to either adapt or move to cooler habitats. Half are already starting to lose that race.

As animals and plants move to higher elevations or away from the equator in search of new homes, their historic ranges shrink, and they go extinct in the areas they leave behind. They've vanished from the local habitats where they've lived for years.

"This is not based on a future projection, it's based on what's already happened," Wiens said. "In some ways, this is just the beginning."[6]

Wiens and others have been studying these "range shifts" now for nearly a decade. In 2008, University of California, Berkeley, biologists Craig Moritz and James Patton studied more than a dozen small mammal species in Yosemite National Park, for instance, and found that they had all moved up to two thousand feet uphill in response to spikes in nighttime temperatures. The alpine chipmunk was being forced to move so far up mountain slopes that it faced total extinction, they found.[7]

Morgan Tingley, a University of Connecticut ecologist, studied birds' ranges in the Sierra Nevada in 2012 and recorded "range shifts" up and down the California mountain ranges by comparing historical data from the 1990s to observational surveys conducted a decade later. He found that birds were moving up and down the mountain slopes due to either changes in temperature or rainfall patterns (which is also a climate signal because between 5 and 10 percent more water vapor is now circulating in the atmosphere). Eventually these species will run out of places they can go, because they can only climb or migrate so far. Plants and animals that can't adapt or move will simply vanish.[8]

Biodiversity researchers are also chronicling the fact that extinction rates are a hundredfold beyond what we should reasonably expect at any given time. Species go extinct all the time. But so many species should not be going extinct as rapidly as they are. That rapid extinction rate means something—that a multitude of factors from environmental pollution and poaching to warming, drought, and extreme heat are making habitats intolerable for many, many more species than is normal.

Researchers from Stanford, UC Berkeley, the University of Florida, and elsewhere released a study in 2015—"Accelerated Modern Human–Induced Species Losses: Entering the Sixth Mass Extinction"—that laid out the problem in stark terms.[9]

"The average rate of vertebrate species loss over the last century is up to 100 times higher than the background rate," the research team wrote in *Science Advances*. In normal circumstances (without human beings creating problems on the planet), the number of species experiencing extinctions at this rate would happen in up to ten thousand years. They occurred in just a century, thanks to us.

"These estimates reveal an exceptionally rapid loss of biodiversity over the last few centuries, indicating that a sixth mass extinction is already under way." Heading this off is still possible, they wrote, "but that window of opportunity is rapidly closing."

Many of Terry Root's biodiversity colleagues met in Australia in 2016 at a big global conference where they could all share information about what they were finding as they studied species migration. It was a sobering conference for everyone who attended. In talk after talk, the same message was repeated over and over: Every species was responding in some fashion to the rapidly changing system on Earth.

Half of all species are now migrating either north or south to the poles to keep from going extinct, one of the world's leading biodiversity experts, Camille Parmesan, told her colleagues in the opening talk of the global conference Species on the Move.[10]

Warming temperatures are causing half of the world's plants and animals to move location. Every major type of species is affected. Parmesan, a professor at Plymouth University, said that data on thousands of species had conclusively found that many had shifted their ranges toward the poles or up mountains since the start of the industrial era.

What's potentially even more profound, she said, is the fact that more than two-thirds of all species across the planet had shifted to earlier spring breeding, migrating, or blooming. Every single type of plant or animal is now being impacted—trees, herbs, butterflies, birds, mammals, amphibians, corals, invertebrates, and fish.

Thousands of biodiversity scientists are studying and chronicling the fact that most of the species and plants on Earth are trying to adapt to something that's happening everywhere presently. And yet there are still people who wonder whether any of it is real or if any of it truly threatens the fabric of life as we know it. This is why Terry Root was so profoundly troubled and why she felt like she'd failed to convince people about the truth of what she and all of her colleagues had found in the past thirty years.

"The global imprint of warming on life is evident in hundreds of scientific studies," Parmesan told her colleagues at the Australia conference. "While about half of all studied species have changed their distributions in response to recent climate change, we are starting to see negative impacts for the most vulnerable species."[11]

Some of the changes are small or subtle. Plants are flowering earlier than they used to. Birds are flying north or south for the winter earlier than they did in previous years. In some cases, she said, species aren't going to make it. "Recovering these vulnerable species under a changing climate may not always be possible," she said.

Scientists who study species on the move and conservation groups are starting to regularly track and report on the most prominent ones—species that vast numbers of people might care about or recognize—because it brings the concept of a sixth mass extinction home.

The long list of iconic species in trouble ranges from the Adélie penguin (which could be decimated within decades), the African elephant (facing extinction as water scarcity shrinks its habitats in Africa), and the American lobster (which may become "Canadian" lobsters as their range shifts from Maine to cooler waters) to the wild turkey (who staged a comeback from overhunting but is now in trouble again as it loses its natural habitat) and the yellow-billed magpie (which could lose its entire winter range within decades).[12]

While Root and her colleagues have always been reluctant to enter the political arena, where questions about the truth of her work tend to become contentious, she hasn't shied away from it. The work is too important. Mass extinctions loom on the horizon, and people need to see what's happening all around them.

"As I look out my window right now I wonder, will the planet look the same if I were sitting here in a hundred years? No. It's not even going to be close," she said in a Union of Concerned Scientists profile.[13] But she also continues to believe that others—like her students who have come through her Stanford classes over the years—will eventually help the world understand what is happening all around us.

"The energy and optimism that the students bring is refreshing and gives me immense hope," she said. "When I have students around that come up to me and say, 'I want to make a difference in the world,' that gives me hope."

3

The Third Pole

The Silk Road, which winds its way from the Mediterranean to the eastern shores of China and the Korean Peninsula, was once the only real trade route linking East and West. It was the best-known route that led through the formidable Himalayan mountain ranges at China's southern border.

One reason that historians write about the Silk Road and trade route travelers like Marco Polo is that they represent a journey through one of the most forbidding environments on the planet. The Himalayas separated East from West for thousands of years. The Silk Road was one of the few points of contact between the two worlds.

The largest mountain ranges in the world cover the southern parts of China. The Himalayas hold the world's tallest peaks along the country's western borders. A combination of towering mountains and treacherous deserts like the Gobi kept early China isolated from Western civilizations for a very long period of time.

While nearly all the Himalaya region has been explored or mapped, there's still a great deal we simply don't know about that part of the world. Routes in and out of the deep recesses of the Himalayas are still confined to single tracks in most places, and very few researchers or scientists have the luxury and wherewithal to study changes there over time.

What we do know about the Himalayas is this: While their towering mountain ranges make it exceedingly difficult to carve out an existence for vast numbers of people, the region itself is almost the sole source of fresh water for more than a billion people who live in countries that exist in the shadow of these great mountains.

There are tens of thousands of glaciers throughout the entire Himalayan mountain range. These glaciers, combined with rainfall, feed into tributaries that cascade down the ranges. They are the source of Asia's largest rivers, including the Yangtze, Yellow, Ganges, Mekong, and Irrawaddy Rivers.

Because the mountainous region contains the largest area of frozen water outside the North and South Poles, it's known colloquially as Earth's Third Pole. The fresh water resulting from that Third Pole supports about 1.3 billion people in China, India, Pakistan, Afghanistan, Nepal, and Bangladesh.

For this reason alone, we'd be right in assuming that scientists would desperately like to study the region and the impacts that have been under way there for several decades. But the truth, sadly, is that the land is so forbidding and harsh that it hasn't been widely studied. We just don't know as much about the rapid changes under way in the Himalayas that are transforming the region as we do other places.

A handful of American, British, and Australian research teams have sporadically studied the glaciers and rising temperatures in the Himalayas over the years. NASA's Earth satellites have studied large-scale glacier melt from the sky. Those satellites can track movements of some of the biggest glaciers from space. What they can't really do, however, is study the thousands of smaller glaciers that have formed in every corner of the series of mountain ranges that make up the Himalayas.

A small number of Chinese researchers, however, have taken up the arduous task of tracking events in certain locations deep in the Himalayas. Until recently, Chinese political leaders didn't make those researchers or their findings from those remote research stations available to Western journalists or scientists on a regular basis.

That's begun to change, especially as China's political leaders have begun to actively engage with Western researchers, nongovernmental organizations (NGOs), and journalists on questions about the ways in which Earth's changing climate is profoundly altering landscapes at remote locations like the Himalayas. China's leaders are now taking Earth's changing climate and its impacts on water and agricultural resources quite seriously.

Chinese research teams at remote stations deep in the Himalayas are starting to report their findings to the outside world. Those findings are surprising . . . and unsettling. For years, scientists have said they didn't expect to see big changes in the Himalayas until the end of this century—long after everyone living today is dead and gone. That rosy assumption appears not to be the case if these Chinese researchers tracking changes at the remote stations are to be believed.

For instance, Chinese authorities opened up a remote research station on the Qinghai-Tibet Plateau to Western journalists. That station has been there for more than a half century. Chinese researchers stationed there have been meticulously taking temperature readings and studying the impact on hundreds of small glaciers in and around the plateau.

ABC Australia's China correspondent, Matthew Carney, and a photographer, Wayne McAllister, filed an extraordinary firsthand report in 2016 from the remote Chinese research station deep in the heart of the Himalayas that has been tracking profound changes in the region virtually unnoticed for decades.[1]

"Deep in the Himalayas sits a remote research station . . . tracking an alarming trend in climate change, with implications that could disrupt the lives of more than 1 billion people and pitch the most populated region of the world into chaos," they reported. "Half a century of research shows the temperature has increased by [2.8 degrees Fahrenheit] in the area, more than double the global average. More than 500 glaciers have completely disappeared, and the biggest ones are retreating rapidly."

Researchers at the Qinghai-Tibet Plateau station have kept daily notes on temperature records and small glaciers for decades. Scientists have known for some time that big glaciers were retreating all over the world in the face of planetary temperatures that have increased, on average, by about 1.4 degrees Fahrenheit since the start of the Industrial Revolution. What this one research station has shown is that temperatures in the Himalayas are double that increase—putting the warmer temperatures at the tipping point for potentially catastrophic impacts—and that hundreds of smaller glaciers have already disappeared. This is happening in the Himalayas right now—not decades from now or at the end of the twenty-first century.

The route to this remote research station passes through the ancient Silk Road town of Dunhuang in northwestern China, at the strategically important junction of the northern and southern Silk Roads. Centuries ago, Dunhuang had a population of seventy-six thousand or so. It was a key supply base for caravans stopping there for water and supplies before attempting the trek across the Gobi Desert. The residents of Dunhuang knew, with their own eyes, that water from the Third Pole was responsible for the oases in various spots in the Gobi. Engineers later learned how to harness water from the glaciers as part of an extensive irrigation system.

Chinese researchers have used the research station located near this important Silk Road crossing for decades in order to study the impacts on the region. It was the very first research station in the Himalayas opened by the Chinese, in 1958, and it has been collecting data ever since. One of China's leading glaciologists, Qin Xiang, told the ABC team that he's seen the changes firsthand there for the past ten years.

The melting glaciers—which are happening much faster than he or anyone else in the Chinese glaciology research community had anticipated—will have a short-term benefit, but that will swiftly give way to devastation as the steady source of water goes away for good. "The volume of water will increase in the short term, but with the shrinking of glaciers in the next thirty years, it will decrease and drastically affect agriculture and life here," he said.

Based on one of the bigger glaciers they track closely near the research station, he also said that the rate of melting has doubled in just the past ten years. It had been steady from the 1960s until the turn of the century. Then it took a giant leap forward and is now accelerating rapidly.

"Based on the figures from 1960 to 2005, in that forty-five years, it only retreated by 260 meters. But in [the past] ten years, it retreated by 140 meters," Qin told the ABC reporting team. "The speed compared to the previous period has nearly doubled."

Based on his own personal observations and reports in that part of the Himalayas from researchers stationed there since 1958, Qin said that the 226 glaciers in that region had lost twenty-seven square kilometers of ice. More broadly, throughout that entire area of the Qilian Mountains (which once held more than 3,000 glaciers), Qin said he'd filed a report that showed 509 smaller glaciers have vanished in the past fifty years. Many more will vanish in the next ten to twenty years.

Because this is, most likely, the oldest continuous monitoring of glacier melting and temperature increases since the late 1950s, the results are important for a host of reasons. "From the data we had over fifty years, it showed in our research areas the temperature increased by [2.8 degrees Fahrenheit]—it is much higher than the national temperature increase," Qin said. "It is because, in the high-altitude areas, the temperature is sensitive to the global warming."

Qin and other research teams have also been monitoring the speed and depth of water flow from the glaciers for fifty years as well. "Compared with the river discharge in 1959, we found the volume of the melting of glaciers has nearly doubled compared to fifty years ago," he said. A combination of black carbon that falls from the sky and settles on the glaciers, combined with rising temperatures, is responsible for the acceleration.

Overall, there are nearly fifty thousand glaciers throughout the Third Pole. These glaciers are critical to the extensive fresh water system that feeds hundreds and hundreds of tributaries descending from the mountains to the deltas and rivers below.

If roughly a sixth of the smaller glaciers in only one region of the Himalayas have already disappeared, and the rates of melting have doubled in the past ten years alone, it's easy to understand why the Chinese are worried about the fate of water supplies throughout its southern region. Others, especially Pakistan and India, should be especially worried as well.

There are big questions, with no easy answers yet, about the implications from this type of research in the Himalayas. Are one-in-a-thousand-year floods now occurring in Pakistan and China linked? What about the accelerating desertification in the region or unprecedented heat waves in India? Global weather patterns like the monsoon season—responsible for fresh water supplies for hundreds of millions of people—are also showing signs of strain, along with the El Niño/La Niña cycles that affect weather events and temperatures. Are these connected as well?

The truth is, we know a great deal about potential regime shifts at the North Pole. We've had research teams at the South Pole (Antarctica) for decades and know quite a bit about the impacts there. We just don't know that much about the Third Pole (the Himalayas), because so much of it is remote, forbidding, and inaccessible. But it's clear from the limited number of studies and amount of research in the region that there are big changes happening there right now. They've been under way for at least a decade. That's why we need to pay special attention to efforts that have tracked real impacts there over time.

Satellite data over the years has shown that there is, or can be, a wide degree of variation in the stability of ice at the top of the world in the Third Pole. Most believe that glaciers are still stable along the Pakistan-China border. But based on both satellite data and firsthand observations like Qin's at the research station near the historic Silk Road way station, this is clearly not the case in all of the Himalayas.

Because the situation is complex, scientists have argued for years about the specific effect of the changes in the Third Pole on water supplies downstream. Changes in the monsoon season, or snowmelt, or groundwater depletion as big cities in China and India grow rapidly, or demands on water from industrial sources all complicate the picture—each contributes to a complicated story of water scarcity.[2]

But researchers who have seen the changes at the Third Pole don't concern themselves with those intellectual or theoretical arguments. They know that the Third Pole is changing much, much faster than other parts of the earth and that the changes will do significant harm in the near future.

Lonnie Thompson, an Ohio State scientist who is perhaps the best-known glaciologist in the United States, has personally visited every major glacier region on Earth during the course of his lifetime of research—including the big glaciers at the Third Pole. Thompson has seen, and studied, them all. He knows the story of glacier melting like no one else alive today.

I watched him once give a closed-door briefing to the National Science Board on his research of the glacier that is disappearing at the top of the famed Mount Kilimanjaro in Africa. NSB members sat stunned as he delivered the sober verdict on the mountain's fate. They may have wished otherwise, but they knew that what Thompson was telling them was true—because he'd been there and observed it firsthand, as scientists routinely do. Studying glaciers isn't theoretical for Thompson. He has personally watched them all melt—and chronicled it for the rest of us in the past forty years or so.

"The summit ice cover of Tanzania's Mount Kilimanjaro . . . has shrunk by 85 percent in the past century," he said after the briefing. "If the current climate conditions persist, it won't be long before Africa's highest mountain is ice-free."[3]

For a long time, Thompson didn't talk publicly about his concerns about what he was witnessing as he studied glaciers. He simply went about his research, reporting his findings to his peers in scientific journals. That all changed a few years ago, when he began to talk more forcibly about how his research had led him to believe that the changes he was observing evidenced a "clear and present danger."

He was the first to tell the world, from firsthand research, that glaciers in Peru that had taken hundreds of years to accumulate had all but vanished in a generation. "At the Quelccaya Ice Cap in the Peruvian Andes, glaciers that accumulated over 1,600 years have melted in just 25 years," he said.

Thompson and a research team traveled to the central Himalayas in 2006 to study the Naimona'nyi Glacier in Tibet. He'd expected to find some evidence of melting. Instead, what they found stunned him—and prompted him to start warning others, he told MSNBC at the time. "At the highest elevations, we're seeing something like an average of [more than half a degree Fahrenheit] warming per decade," Thompson said.[4] Some of the smaller glaciers he visited had melted away, exposing rock from the time of the Second World War.

Thompson is on a mission to warn the world of what's coming. Almost nothing stops him—not even a heart transplant in 2012. Three years after receiving a new heart, he joined a research team that hiked to the top of the Guliya Ice Cap in western Tibet in the fall of 2015. It was the highest altitude a heart transplant recipient had ever reached.

Thompson has no use for those who deny reality, for whatever reason. He has personally studied glaciers for his entire professional career. He has also risked his life for research about what the fate of glaciers at the Third Pole and elsewhere mean for all of us on the planet. He simply wants others to understand what he's witnessed and observed at the top of every high mountain range on Earth. He has led more than sixty research expeditions around the world, but the fate of the Third Pole has always been a special obsession for him, because it is so intricately linked to impacts that could affect hundreds of millions of people shortly.

What happens at the Third Pole is, perhaps, more important even than changes that are occurring in the Arctic or Antarctica, he says, because they have such a dramatic impact on big weather phenomena like the monsoon season or El Niño.

"What happens [at the Third Pole] is extremely important for understanding our changing climate, especially phenomena such as monsoons and the El Niño–Southern Oscillation," Thompson told *Scientific American* after his most recent research expedition to Tibet. "This is where we have the highest sea-surface temperatures and where large quantities of water vapor are produced by evaporation from the oceans. They are the engine of the climate system."[5]

What he has consistently found, and reported meticulously in scientific journal studies, is the reverse of what people might otherwise assume—namely, that temperatures actually increase (compared to historical norms) the higher the elevation. This is the crux of the problem, and the threat, facing the Himalayas.

"The higher the elevation, the greater warming we have," he said. "This is in line with the observation that the vast majority of glaciers in Tibet and the Himalayas are retreating. In some extreme cases, as ice cores from the Naimona'nyi Glacier in southern Tibet show, all the snow and ice that accumulated since 1950 has melted or sublimated away at altitudes as high as six thousand meters above sea level."

The rapidly changing ecosystem of the Third Pole—even more so than the changes presently under way in the Arctic—could set nation against nation if we don't soon come to grips with the reality of the changes occurring there. This is the story that Thompson and other glaciologists have tried to tell us for some time. The Third Pole is at the very center of those stories they've struggled to tell.

The good news is that research teams from every continent recognize the urgent imperative to share knowledge in hard-to-study regions like the Third Pole. They now broadly share research findings and consistently work together on joint research expeditions. The researchers' hope is that political leaders recognize the need to share knowledge—and what it means— as they and their peers have.

4

Collapse of the Pollinators

A few years ago, bee enthusiasts, agricultural scientists, plant gene-ticists, and conservation societies started talking. The rusty-patched bumblebee—endemic to North America and once the most common, rec-ognizable type of bumblebee in much of America—was in trouble. With its distinctive black-and-gold features (hence its name), this bumblebee is easy for people to spot as they watch it pollinate flowers and crops in the field. But its numbers had dropped precipitously from a combination of factors that included pesticide use and the changing climate system that disrupted its narrow habitat range.

By 2016, the numbers were too hard to ignore. The bumblebee that most schoolchildren in America associate with bumblebees more broadly had all but vanished in locales across North America. The U.S. Fish and Wildlife Service put the rusty-patched bumblebee on the endangered species list in January 2017. Conservation and science societies now talk about the rusty-patched bumblebee in the past tense.

"This bee was once commonly distributed throughout the . . . United States, but has declined from an estimated 87% of its historic range in re-cent years," says the Xerces Society, a wildlife conservation NGO that first filed the petition to list the bumblebee as endangered in 2013. "The rusty-patched bumble bee was once an excellent pollinator of wildflowers, cranberries, and other important crops, including plum, apple, alfalfa and onion seed."[1]

There are thousands of bee species around the world . . . 3,600 different species in North America alone. Most people think that bees exist to make honey. But virtually none of the bee species in North America do this—the vast majority are pollinators that make flowers and crops grow.

Bumblebees are a relatively small group of bees. There are 250 species of bumblebees in the world and 47 in North America. Their physical characteristics (larger body size, denser hair) make them especially sensitive to changes in the climate. They're active in cooler temperatures and can live in higher elevations. For this reason, when temperatures get hotter, they struggle to survive. The rusty-patched bumblebee ran precisely into this wicked dilemma, says the Xerces Society.

Bumblebees are especially good as "buzz pollinators" (where they vibrate with their wings while clinging to the flower, causing pollen to dislodge). It's why they are the primary pollinators for important agricultural crops like greenhouse tomatoes, for instance.

The rusty-patched bumblebee, sadly, isn't alone. Other bees are in trouble as well.

"In the late 1990s and early 2000s while conducting studies of Franklin's bumble bee (*Bombus franklini*) and the western bumble bee (*Bombus occidentalis*) near Mt. Ashland in southern Oregon, Dr. Robbin Thorp, Professor Emeritus of entomology at University of California, Davis, observed a sudden and dramatic decline in the individuals of these species," Xerces says in a special report on the plight of the bumblebee in America.[2]

Since his discovery, Thorp has been one of the leading scientists striving to understand what is happening to bumblebees all across North America and what is driving the collapse. They have partial answers now.

"While the causes of the declines of the rusty-patched bumble bee and other North American bumble bee species are not entirely understood, factors contributing to declines likely include pathogens amplified by commercial bumble bees, habitat loss, pesticide use, and climate change," Xerces concludes in its report.

Some bumblebees are bred to pollinate in commercial settings. When they interact with bumblebees in the wild, it seems to cause problems. Pesticides are also undoubtedly causing harm. So is development and land use, both of which destroy natural bumblebee habitats. All of this, combined with the changing climate system as a whole, has created a deadly brew forcing the rusty-patched bumblebee and others like it toward the edge of extinction.

Bees are especially sensitive to changes in their narrow habitat. Earth's changing system is a particularly difficult proposition for them—especially for the rusty-patched bumblebee. A groundbreaking study in *Science* in 2015 (similar in scope and ambition to Terry Root's seminal bird-migration study earlier) showed just how fragile the situation was for bees overall.[3]

"As the climate changes, plants and animals are on the move," *Science* magazine's Cally Carswell wrote about the study. "So far, many are redistributing in a similar pattern: As habitat that was once too cold warms up, species are expanding their ranges toward the poles, whereas boundaries closer to the equator have remained more static. Bumblebees, however, appear to be a disturbing exception."

Researchers looked at dozens of bee species in North America and Europe and found that they were failing utterly at moving at the same pace as temperature rise. They were simply unable to colonize new habitats north of their historic range and were disappearing from the southern portions of their habitat range.

"Climate change is crushing [bumblebee] species in a vise," said Jeremy Kerr, an ecologist at the University of Ottawa who was the study's lead author.

The study was massive. It compiled a data set that consisted of more than four hundred thousand direct observations of bee colonies dating all the way back to 1901. The researchers were able to analyze the movements of sixty-seven bumblebee species in North America and Europe by working through this massive data set of observations.

From there, they then mapped out the changes in territories for the dozens of bee species in their data set. The map laid out their "thermal ranges"—from the warmest to the coolest places that the bees lived. They included a variety of factors that might have explained the changes, ranging from the changing Earth system to land cover and pesticide use.

Like Root's historic bird-migration study, the bee researchers found that the dozens of bee species observed for more than a century had retreated by nearly two hundred miles from the southern edge of their historic ranges since 1974.

The rusty-patched bumblebee was perhaps hardest hit of all, they found. It had disappeared entirely from the southeastern United States.

"Southern species (of bees) are also retreating to higher elevations, shifting upward by an average of about 300 meters over the same time period," Carswell wrote. "Meanwhile, few species have expanded their northern territories."

The study conclusively showed that the changing climate system in North America and Europe was the only factor that had a meaningful impact on the large-scale range shifts.

Why were researchers so confident that the bees were moving due to the changes under way to Earth's system? Because the observations on the massive shift began well before neonicotinoid pesticides (which many also blame for threats to the bees) came into widespread use. The retreat from southern territories happened quite quickly—and was almost certainly in response to an increase in temperatures that the bees simply weren't prepared to deal with.

While the plight of the rusty-patched bumblebee may seem confined to the narrow world of conservationists and bee hobbyists, it's far from that. A disturbingly similar story is playing out elsewhere in the world with thousands of other types of pollinators. The rusty-patched bumblebee is a cautionary tale that has galvanized leading scientists in agriculture, plant genetics, and environmental science on a worldwide basis in ways never seen before.

Dozens of scientists who study pollinators and their role in global agriculture are so profoundly worried, in fact, that they pushed world leaders to organize a global committee of scientists to assess the threats to pollinators everywhere. Eighty scientists offered to support the effort and promised to spend years of their lives on a special United Nations report on pollinators.

The first assessment of pollinators—part of a broader biodiversity effort called the Intergovernmental Science-Policy Platform on Biodiversity and Ecosystem Services that included scientists from more than one hundred countries—was issued in early 2016. Their initial findings were sobering.

"Without an international effort . . . increasing numbers of species that promote the growth of hundreds of billions of dollars' worth of food each year face extinction," science journalist John Schwartz reported in *The New York Times*.[4]

The report was the first global assessment of threats to pollinators, which included twenty thousand species of wild bees along with birds, butterflies, moths, wasps, beetles, and bats. The scientists promised to continue to assess threats to pollinators and report their findings on a regular basis for world leaders because of the pollinators' critical role to the food system and the worldwide agriculture economy.

Pollination—the transfer of pollen between the male and female parts of flowers to enable fertilization and reproduction—is critical to much of the world's food production. The majority of cultivated and wild plants depend in some measure on a variety of animals to transfer pollen.

More than three-quarters of the leading types of global food crops rely to some extent on pollination (either for yield or quality). Pollinator-dependent crops make up a third of all global crop production volume. Over the past fifty years, food crops that rely on pollination have increased 300 percent in order to feed the world's growing population. The annual value of that crop production volume is nearly $600 billion. It's responsible for millions of jobs.

"Pollinators, which are economically and socially important, are increasingly under threat from human activities, including climate change, with observed decreases in the abundance and diversity of wild pollinators," the authors wrote in a summary of the initial assessment. The group didn't conduct new research. Instead, it looked at every relevant science study on pollinators in recent years in order to formulate its first worldwide assessment of the threat.[5]

As dozens of scientists from 124 countries compared notes on thousands of studies, what they found alarmed them. Many pollinator species are threatened with extinction right now. Roughly two of five invertebrate pollinators (like bees and butterflies) are on a path toward extinction. The list of pollinators threatened with extinction includes 16 percent of vertebrates (e.g., birds and bats). That overall list of vertebrate species includes iconic hummingbird species and hundreds of other bird species that feed on nectar and spread pollen from one flower to the next.[6]

Like the rusty-patched bumblebee in America, the threats to pollinators around the world are complex and varied. In some cases, pollinators are in trouble in the face of aggressive commercial agriculture practices that destroy natural habitats. In other cases, parasites introduced in changing ecosystems are a problem. Pesticides that introduce pathogens are a problem as well.

In all cases, the changing climate weaves in and out of the thousands of ecosystems as a constant threat, creating a whipsaw effect for species. As temperatures rise everywhere and planting and flowering times change, it affects the territories of both plants and pollinators. In many instances now, there's a mismatch. The pollinators simply don't know where—or when—to find their partners.

"Will the pollinators be there when the flowers need them?" asked Bob Watson, who is a vice chair of the UN assessment of pollinators and a senior official at the Tyndall Centre for Climate Change Research.[7]

Several of the scientists involved say publicly that it will only get worse from here on out, even if government and business leaders act quickly. "There are going to be increasing consequences," one of the report's lead authors, Simon Potts, told the Associated Press's national science reporter Seth Borenstein.[8]

As we've seen, there are biodiversity threats everywhere in the world now, with local extinction rates at 50 percent in one snapshot of a thousand different species. The extinction rate overall in the world now is a hundred times what would be expected based on historical norms.

But it was the threat to pollinators, which play such a critical role to so much of the world's agricultural process, that leading biodiversity experts wanted to focus on first in a series of threats it plans to assess in coming years.

"The report is the result of more than two years of work by scientists across the globe who got together under several different U.N. agencies to come up with an assessment of Earth's biodiversity, starting with the pollinators," Borenstein wrote. "It's an effort similar to what the United Nations has done with global warming, putting together an encyclopedic report to tell world leaders what's happening and give them options for what can be done."

The scientists who study pollinators say that people need to pay close attention to what's happening now. "Everything falls apart if you take pollinators out of the game," said University of Maryland bee expert Dennis vanEngelsdorp. "If we want to say we can feed the world . . . pollinators are going to be part of that."

Similar to ways that the Sahara Desert can be transformed in the Sahel in Africa in order to turn back the clock, changes can be made locally that can also have an immediate impact on threats to pollinators right now.

Much as local farmers in Niger and Burkina Faso have discovered ways to keep the deadly combination of water drought and extreme rainfall events from denuding land by planting trees with crops and developing indigenous ways to capture and store water, local farmers can likewise make similar moves to bring the pollinators back from the brink of extinction before it's too late.

"There are relatively simple, relatively inexpensive mechanisms for turning the trend around for native pollinators," said David Inouye of the University of Maryland, a coauthor of a couple of chapters in the report.

Like the Niger and Burkina Faso examples, diversity is the key. One of the biggest problems is that pollinators historically do much better in diverse environments (like grasslands that are littered with wildflowers, for example).

Yet large agribusinesses tend to create huge plots of farmland dedicated to just one crop. There are no wildflowers spread throughout them. Europe—where 97 percent of grasslands have disappeared since the Second World War—is especially bad for pollinators this way. So is the United States.

At least one country—England—has started to experiment with diversity. It now pays farmers to plant wildflowers within crops, and the experiment seems to be working.

Other experiments like organic farming and planting flower strips also seem to hold promise. Where the experts seem to agree is in the need to avoid big swaths of farmland that have stripped out diversity—because that lack of crop, flower, or plant diversity runs counter to what pollinators have historically flourished in over the centuries.[9]

The biodiversity experts who put together the pollinators report have learned from experts in the Sahel and elsewhere where academics and experts have studied how and why indigenous practices have obvious benefits. Diversification, local knowledge, and indigenous farming practices are front and center in the pollinators report. Its most important recommendations zero in on this concept.

The best hope for pollinators lies in "strengthening existing diversified farming systems (including forest gardens, home gardens, agroforestry and mixed cropping and livestock systems) to foster pollinators and pollination through practices validated by science or indigenous and local knowledge (e.g., crop rotation)," the authors of the UN report concluded. Restoring or protecting diversified patches of natural habitats throughout farming landscapes is also critical.

Simple things, such as letting local farmers know when flowers and crops might need earlier pollination as the climate system changes, can be handled by education through social media and radio. Earlier seasonal indicators (like flowers blooming) can be communicated to local farmers, in turn triggering decisions to plant crops that require pollination.

Similarly—and a bit surprisingly—weeds tend to mix things up for wild pollinators and help them build up immunity to other threats in their environment. Herbicides that kill off everything but the one crop they're designed to benefit may have an unanticipated side effect of harming the pollinators needed for those crops to thrive. In short, bees love weeds.

"Pesticides, particularly insecticides, have been demonstrated to have a broad range of lethal and sub-lethal effects on pollinators in controlled experimental conditions," the UN biodiversity report said. But they also kill off weeds, which can be useful for wild pollinators, so the scientists urged local farmers and agriculture businesses to look hard at ways in which they can diversify flowers and plants in fields where pollinators are needed.

Jeremy Kerr, the Ottawa researcher who led the team that studied a century's worth of observations of sixty-seven bee species and discovered that they were all trying to move north to adapt to a rapidly changing Earth system, said that bees are a perfect illustration of what we're confronting across the board with pollinators.

The bees are also dealing with threats coming at them from all directions in modern society in the twenty-first century. "We're hitting these animals with everything," he told *Science*. "There's no way you can nail a bee with neonicotinoids, invasive pathogens, and climate change and come out with a happy bee."[10]

The loss of bee species and other pollinators will have enormous consequences that we're only now coming to terms with. "We play with these things at our peril," Kerr said. "The human enterprise is the top floor in a really big scaffold. What we're doing is reaching out and knocking out the supports."

5

The "Evil Twin"

Scientists occasionally refer to ocean acidification as the "evil twin" of climate change. Just as we can't truly see the black sky overhead that's causing disturbances to Earth's system on a mass scale, we can't truly see the changes that have happened in our oceans as carbon dioxide has dissolved in seawater since the start of the Industrial Revolution.

But the changes are there—and they are now threatening to harm thousands of species on a scale that human beings have never seen before. The ocean's chemistry is currently changing faster than at any time in the recent past.

A significant percentage of the carbon that's released when we burn coal, oil, and gas doesn't hover above us in the atmosphere. It's dissolved and stored in oceans that cover most of the surface of the earth. For a long time, scientists believed that this was actually a good thing because it removed a substantial portion of the carbon that would otherwise be trapped in the atmosphere above us that heats the planet.

That belief, like the wishful thinking that we wouldn't see mass coral reef die-offs until much later in the twenty-first century, seems quaint today.

We are now pouring 22 million tons of carbon into the oceans each and every day. The oceans have absorbed 525 billion tons of CO_2 since the beginning of the industrial era. Because carbon dioxide dissolves in seawater, it changes the ocean's chemistry.

Every high school chemistry class in America has studied some version of this type of experiment: What happens when CO_2 is dissolved in water? The water becomes more acidic. What scientists had always hoped was that the vastness of the oceans could absorb all this carbon without any significant impact on life and ecosystems.

That hasn't been the case. Today, the world's oceans are 30 percent more acidic than they were at the start of the industrial era. This has profound implications—for marine life, for the health of ocean ecosystems, and for us.

"Even though the ocean is immense, enough carbon dioxide can have a major impact. In the past 200 years alone, ocean water has become 30 percent more acidic—faster than any known change in ocean chemistry in the last 50 million years," according to a special report on the growing impacts of ocean acidification by the Smithsonian's National Museum of Natural History.[1]

When scientists first started studying whether it was even possible for oceans to grow more acidic over time, they initially assumed that the millions of rivers that carried fresh water into oceans in every corner of the earth would "balance" out the dissolved carbon and stabilize the pH in the ocean.

In theory, this made sense. The problem, they discovered over time, was that Earth's system was changing much too rapidly. Far too much carbon was being dissolved in the ocean. Not only was it accumulating in the sky above us, it was overwhelming the oceans as well.

"So much carbon dioxide is dissolving into the ocean so quickly that this natural buffering hasn't been able to keep up, resulting in relatively rapidly dropping pH in surface waters. As those surface layers gradually mix into deep water, the entire ocean is affected," the Smithsonian report concluded.

This very rapid change is also a massive shock to the system for marine life. They simply haven't had enough time to adapt.

"Such a relatively quick change in ocean chemistry doesn't give marine life, which evolved over millions of years in an ocean with a generally stable pH, much time to adapt. In fact, the shells of some animals are already dissolving in the more acidic seawater, and that's just one way that acidification may affect ocean life," according to the Smithsonian report.

The impacts on ocean life are only just now becoming known. While scientists began to study ocean chemistry thirty years ago, they only started to look at potential impacts on ocean life a decade or so ago. The first biological impact study on ocean life began in 2003, when scientists first started to recognize that the phenomenon of acidification was even possible in such vast bodies of water.

Not surprisingly, given that we have only just begun to study the problem, the immediate future for ocean species is largely unknown at this point. Some species will adapt or move as the water grows more acidic. Others will go extinct. But one thing is certain: The ocean's biodiversity is now being affected, and it will impact communities that survive from this biodiversity just as they survive on fish in and around coral reefs.

There's an important principle to understand here. A little bit of carbon dioxide is actually helpful on land. Plants need it to grow. Human beings and animals alike exhale carbon dioxide. But we are now adding CO_2 to the atmosphere faster than we should be, which is overwhelming Earth's system.

A similar problem, in reverse, is occurring in the oceans. When CO_2 dissolves in seawater and the pH levels are lowered, a series of chemical changes occurs that breaks down CO_2 molecules and recombines them with others. The additional CO_2 doesn't simply float around in the seawater. The dissolving process binds carbonate ions (which makes them less prevalent in the water) that marine creatures need to build shells and skeletons. Calcium carbonate minerals are the building blocks for the skeletons and shells of many marine organisms. This is what is threatening marine life, and scientists are only just now studying it in myriad ways.[2]

Chemistry classes everywhere study this effect. When water (H_2O) and CO_2 mix, it forms carbonic acid (H_2CO_3). While this sort of acid is considerably weaker than other forms of acid (like sulfuric acid in car batteries or hydrochloric acid in our stomachs), it still works in the same way as any acid. It also releases hydrogen ions and binds carbonate ions that creatures with shells need to live.

While the oceans are big and the changes so far may seem small, they're not. "So far, ocean pH has dropped from 8.2 to 8.1 since the Industrial Revolution, and is expected to fall another 0.3 to 0.4 pH units," the Smithsonian report said. "If we continue to add carbon dioxide at current rates, seawater pH may drop another 120 percent by the end of this century, to 7.8 or 7.7, creating an ocean more acidic than any seen for the past 20 million years or more."[3]

Small changes in pH levels mean something. Doctors look at changes in pH levels in blood to gauge whether something is going wrong in our bodies. A drop in our blood pH levels of just 0.2 (the same sorts of pH drops we are now seeing in oceans) can cause seizures, comas, and death in humans. Changes in pH levels in the ocean mean much the same thing.

Here's what worries marine scientists now that they've begun to study the change in the ocean's chemistry. The last time the oceans experienced these same types of changes in chemistry that we're seeing today, it was enough to trigger a mass extinction event. Even the deep-sea invertebrates vanished in that mass extinction event in the oceans fifty-five million years ago.

Is a 30 percent increase in ocean acidification today comparable to what triggered that mass extinction event fifty-five million years ago? We don't know, but marine scientists are racing to find out because the stakes are enormous. More than one billion people rely on the sea for their food. Half of those are threatened at present by the collapse of the coral reef systems evident in the death of so much of the Great Barrier Reef and other reefs. The other half rely on shelled creatures and other marine life that are at risk from acidifying seawater.

"When shelled organisms are at risk, the entire food web may also be at risk," said a special report on the issue from NOAA's carbon program. "Today, more than a billion people worldwide rely on food from the ocean as their primary source of protein. Many jobs and economies in the U.S. and around the world depend on the fish and shellfish in our oceans."[4]

A 2013 study by researchers at the University of California, Santa Cruz, found that acidification was, in fact, reducing the density of coral skeletons and making them more susceptible to disruption. It was the first study to show that coral reefs weren't going to be able to adapt to the rapidly changing ocean chemistry.[5]

The UC team studied coral reefs along the Caribbean coastline of Mexico's Yucatan Peninsula (where submarine springs lower the pH of the surrounding seawater in a localized, natural setting) in order to simulate the global ocean acidification process. The study, published in *PNAS*, showed that there were likely to be major shifts in coral reefs.

But it is species with shells that are certain to be harmed the most. A hard shell is the best defense for creatures in the ocean that might otherwise be easy prey—and acidification studies are starting to show widespread impacts on shelled creatures in seawater. Some are capable of adapting, but many are not.[6]

Some unusual impacts are also being revealed by other studies. Mussels, for instance, can't cling as much as they used to on the bottom of boats or on rocks, one study found.

"The shellfish industry, already adjusting to the fact that acidifying oceans hurt the abilities of sea creatures like oysters to make their shells, is also likely to experience losses when mussels lose their ability to cling," *Scientific American* reported. "Worldwide, mussel cultivation is a major food source and a $1.5 billion industry, according to the U.N. Food and Agriculture Organization. In Belgium, for example, catching and eating mussels is almost a national pastime."[7]

Other studies show that more acidic waters likely impact some marine life at the initial stages of their creation. It was probably responsible for a mass die-off of oysters off the coasts of Oregon and Washington in recent years.

From 2006 to 2008, oyster larvae off Oregon's coast began dying dramatically. Oyster farmers there sent samples to labs—and discovered that acidification was the likely culprit. Other studies confirmed the same problem.

"For the past six years, wild oysters in Willapa Bay, Washington, have failed to reproduce successfully because corrosive waters have prevented oyster larvae from forming shells," Elizabeth Grossman wrote in *Yale Environment 360*.

"Wild oysters in Puget Sound and off the east coast of Vancouver Island also have experienced reproductive failure because of acidic waters. Other wild oyster beds in the Pacific Northwest have sustained losses in recent years at the same time that scientists have been measuring alarmingly corrosive water along the Pacific coast."[8]

The NOAA special report also focused on this problem, which threatens a $100 million industry in the United States alone. It's too early to say, definitively, whether ocean acidification is primarily responsible. Scientists have only begun to unravel the mystery behind it. But the signs clearly point in this direction.

"In recent years, there have been near total failures of developing oysters in both aquaculture facilities and natural ecosystems on the West Coast," the NOAA said. "It is premature to conclude that acidification is responsible for the recent oyster failures, but acidification is a potential factor in the current crisis to this $100 million a year industry."[9]

Another study showed that sea butterflies (pteropods) were dissolving in the Southern Ocean, where more acidic waters tend to rise to the surface from deep ocean currents.[10]

An unusual fear that has emerged more recently is that jellyfish will continue to thrive in more acidic waters—a development that could become deadly if jellyfish compete with fish and other predators for food and disrupt the entire ocean ecosystem. If jellyfish dominate these ecosystems, the consequences could be catastrophic.[11]

Early studies also reveal that acidification is truly the "evil twin" of climate change for coral reefs. Several large-scale studies have shown that coral reefs—which are severely threatened and dying in warm waters—may also be disrupted in the early stages of their creation and regrowth as they try to recover.

Ocean acidification "has the potential to impact multiple, sequential early life history stages, thereby severely compromising sexual recruitment and the ability of coral reefs to recover from disturbance," researchers concluded in a seminal paper published in *PNAS*.[12]

The premier science body in America—the National Academies of Sciences, Engineering, and Medicine—summarized the current state of the science of ocean acidification, which is still evolving because it is such a new field. There will be "winners and losers," NAS said in its definitive report published by the National Academies Press.[13]

"Ocean acidification has the potential to disturb marine ecosystems through a variety of pathways," it said. "[This] will result in ecological winners and losers, as well as temporal and spatial shifts in interactions between species . . . leading to changes in predator-prey, competitive, and other food web interactions."

There is one area that has marine scientists especially worried—the impact on phytoplankton, which drives the entire food web in the oceans.

"Many of the physiological changes from ocean acidification are expected to affect key functional groups—species or groups of organisms that play a disproportionately important role in ecosystems," the NAS reported. "These include expected effects on phytoplankton, which serve as the base of marine food webs, and on ecosystem engineers, which create or modify habitat (e.g., corals, oysters, and seagrasses). Such changes may lead to wholesale shifts in the composition, structure, and function of these systems and ultimately affect the goods and services provided to society."

Here's why this matters—and why it has scientists so worried. Phytoplankton is the primary source of food for much of the marine life that exists in the oceans. Scientists already know that unrestrained phytoplankton growth (known as algae blooms) can have deadly, toxic effects. Anything that changes the beginning of the marine life food system—as acidification appears to do at some level on phytoplankton—will quite literally change life as we know it in the oceans.

"In a balanced ecosystem, phytoplankton provide food for a wide range of sea creatures including whales, shrimp, snails, and jellyfish," says the National Ocean Service. "When too many nutrients are available, phytoplankton may grow out of control and form harmful algal blooms. These blooms can produce extremely toxic compounds that have harmful effects on fish, shellfish, mammals, birds, and even people."[14]

When a research team at MIT studied the potential effects of acidification on phytoplankton two years ago, the results took the research team by surprise. The research was a comprehensive meta-analysis of 154 different papers on phytoplankton published in 49 peer-reviewed journals. The lead MIT scientist on the project said she was "shocked" at the study's implications for the basis of marine life in the oceans.

"Increased ocean acidification will dramatically affect global populations of phytoplankton—microorganisms on the ocean surface that make up the base of the marine food chain," wrote MIT science writer Jennifer Chu.[15]

The MIT researchers concluded that substantial changes to phytoplankton will have profound impacts on virtually every species in the ocean. Some will die out. Others might survive largely untouched. But every species will be affected.

"The researchers report that increased ocean acidification . . . will spur a range of responses in phytoplankton," Chu wrote. "Some species will die out, while others will flourish, changing the balance of plankton species around the world."

The principal research scientist at MIT's Center for Global Change Science, Stephanie Dutkiewicz, said the phytoplankton study startled her because it suggested that an upheaval of the very basis of the entire ocean food web was currently under way.

"I try not to be an alarmist, because it's not good for anyone," Dutkiewicz said when the study was published. "But I was actually quite shocked by the results. The fact that there are so many different possible changes, that different phytoplankton respond differently, means there might be some quite traumatic changes in (all marine life) communities."

Her conclusion from the comprehensive study is that acidification's effects on phytoplankton will change everything, whether we like it or not. "The whole food chain is going to be different," she said.

Marine scientists are also intently focused on the potential for regime shifts in oceans—a tipping point or threshold beyond which the entire food web shifts to an entirely different ecosystem. Studies in recent years have shown that such regime shifts in an ocean ecosystem can occur quickly and without warning. Such a regime shift in the oceans would likewise produce significant "winners and losers"—while also threatening communities that derive either their livelihood or their food supply from a stable ocean environment.

"Regime shifts are likely within those marine ecosystems that experience stress from ocean acidification . . . particularly in combination with other stressors," according to the NAS report. "Ecosystems degraded by acidification also may become more sensitive to other human and climate change stressors beyond ocean acidification.

"Analyses of previous regime shifts in both terrestrial and marine ecosystems, lakes, coral reefs, [and] open ocean show that they were rarely predicted, and many appeared to be triggered by relatively small events," it concluded.[16]

As we see in other parts of the world, big changes in the oceans are due to a variety of factors. Drought and water scarcity in the subtropics, for instance, is occurring from several directions at once. The oceans—and threats to marine life—are much the same.

"A major challenge in ocean acidification research is how to attribute ecological shifts to forcing from ocean acidification. In the field, ocean acidification rarely, if ever, will be the only driver of change," the NAS said. "Climate change is simultaneously causing changes in temperature, circulation patterns, and other phenomena, so that attribution of changes (or at least part of the change) to ocean acidification will be difficult."

Regardless of which factors are more important than others, marine scientists are closing in on some troubling conclusions. Marine life as we know it may be perilously close to regime shifts as we learn more about the effects of ocean acidity. Phytoplankton research is an example of such a driving concern.

These changes, in turn, are certain to have an enormous impact on more than a billion people whose livelihoods and communities depend on a stable marine life ecosystem.

"Marine ecosystems provide humans with a broad range of goods and services, including seafood and natural products, nutrient cycling, protection from coastal flooding and erosion, recreational opportunities," the NAS report concluded. "Many of these goods and services may be affected by ocean acidification."[17]

Finally, in a field that has seen only limited research, some scientists are starting to worry about the effect of changes on phytoplankton to the oxygen that we need for life itself on Earth. Ocean phytoplankton—through the photosynthesis process—is responsible for as much as 50 percent of the oxygen that we breathe. At least one study has tracked a 6 percent drop in phytoplankton levels in just the past thirty years. The question is whether such a precipitous drop might also affect oxygen levels.

"The concept of changes to ocean productivity and ecosystems due to acidification is a very important one to consider," Ken Buesseler of Woods Hole Oceanographic Institution in Woods Hole, Massachusetts, told *Discovery News*. "If half of the photosynthesis on the planet is in the ocean and if you reduce that because of acidification, that is a big deal."[18]

To put it simply, because plants produce oxygen and recycle carbon dioxide through the process of photosynthesis, fewer plants means less oxygen to breathe and less recycling of carbon dioxide. And since ocean plants also feed marine life, which eventually ends up feeding other animal and human life, acidification greatly affects the sustainability of life on our planet.

Ocean acidification, species extinction and migration, the melting of the Third Pole, and the loss of pollinators are all harbingers of Earth's changing climate. Though thousands of modern-day Einsteins are yelling as loudly as they can about the threats, it will be up to us whether we believe those threats and choose to take them seriously.

PART 2

The Ecosystems

Large, unprecedented regime shifts in our atmosphere, in our oceans, and at the ends of the earth are only beginning.

What happens globally when the Arctic breaks long-term weather patterns and is warmer than normal? When the Sahara Desert spreads farther south each year, threatening to swallow all of North Africa whole? Or when a section of the Great Barrier Reef dies due to ocean system shifts? All three are harbingers of a much bigger, cataclysmic story of climate change on planet Earth.

"The Ecosystems" explains how global warming has created large regime shifts, irreversibly changing landscapes across the world as we know them.

What happens in the Arctic disrupts nearly all weather patterns across the Northern Hemisphere, may permanently alter the jet stream, and will inevitably lead to geopolitical regime shifts globally, with so many economically interested parties converging in the Arctic.

With tens of thousands of farms in the Sahel being swallowed by the Sahara Desert, what will happen to Africa's arable land, food reserves, and workforce ten years from now? How will such changes affect the numbers of climate refugees globally? Drive military conflicts?

With coral reefs dying off, what will happen to the large diversity of marine life they support in a symbiotic relationship? To the fish that feed millions around the world?

The problems and questions are undeniably huge, but addressing the fate of such locales as the Arctic, the Sahel, and the Great Barrier Reef and finding solutions now will impact not only the future of every plant, animal, or human living in those particular regions but ensure our survival on an interconnected planet.

6

Regime Shift

An extraordinary thing happened in November 2016. It was thirty-seven degrees Fahrenheit warmer in the Arctic than it usually is in that part of the world during the winter. Those who study the Arctic or live there—from meteorologists to indigenous villagers who've grown accustomed to rapid changes in that region—were shocked.

What was even more extraordinary, and made the extreme situation quite real to people all over North America, is that the usual cold air blasts coming down from the north were strangely absent that year. It was seventy-two degrees on Christmas Eve in New York City. It's never been that hot in New York during Christmas—ever.

"The Arctic is broken," Paul Huttner, the chief meteorologist for Minnesota Public Radio, wrote in an attempt to explain why temperatures in the Arctic had migrated south to North America, producing record-breaking temperatures everywhere in the United States. Taken as a whole, temperatures across North America were as much as twenty degrees higher than normal at the end of 2016.[1]

For scientists and others studying the long-term trends in the Arctic, and the impact that global warming has on both the ecosystem and the natural resources in the area, the temperatures during the winter of 2016 were deeply troubling.

"It's been about 20C [37 degrees Fahrenheit] warmer than normal over most of the Arctic Ocean, along with cold anomalies of about the same magnitude over north-central Asia. This is unprecedented for November," said Rutgers atmospheric science professor Jennifer Francis.[2]

"These temperatures are literally off the charts for where they should be at this time of year. It is pretty shocking. The Arctic has been breaking records all year. It is exciting but also scary," said Francis, who has been studying ways in which the extreme temperature shifts in the Arctic in recent years are starting to have a dramatic effect on weather patterns in North America and Europe.

Sea ice, which forms and melts each year, has declined more than 30 percent in the past twenty-five years. It hit the lowest levels in recorded history the same month that temperatures spiked so dramatically in the region. Francis is unambiguous about the reason—and what it means. "It's all expected," she said. "There is nothing but climate change that can cause these trends. This is all headed in the same direction and picking up speed."

Scientists have several possible explanations for what's happening in the Arctic, ranging from the aftereffects of El Niño to the way in which heat escaping from the Arctic Ocean is still lingering in the atmosphere. What happened in the Arctic in the winter of 2016 could be a pattern or simply an outlier. We may never see extreme temperatures like this again in that part of the world—or we may start to see them quite often.

Francis has researched the extraordinary way in which extreme temperatures in the Arctic have likely changed the jet stream—a process that may make wild temperature swings in North America more common and accelerate ice sheet melting in places like Greenland.

She has studied this issue for years, often in pitched battles with other scientists who have begun to gradually come around to her hypothesis. Her thesis is that rising temperatures in the Arctic, closely intertwined with the loss of sea ice, are changing the shape of the jet stream and altering the weather of the Northern Hemisphere.

"We've been watching the Arctic very closely for years, but [2016 has been] particularly interesting," Francis told Yale's *E360* news service. "It's not just the bizarre behavior that's going on right now, but really it started in the beginning of this year. It's been pretty much in record territory in terms of sea ice and high temperatures almost every month.

"Of course, this is continuing a trend that has been going on for a few decades now where we've been watching the sea ice decrease, mostly in the summer, but also in the other parts of the year," she added. "We see the Arctic warming much faster than elsewhere on the globe. It's really part of a bigger, longer story that has unfolded in a big way this year. What we're seeing happening now is completely different from what we would have seen if we could have looked down on the earth even 30 years ago."[3]

What Francis and other scientists have been studying in the Arctic for years now (and trying to warn the rest of us about) is this: Temperature increases in the Arctic now are two or even three times greater than other parts of the world. Besides melting sea ice—the volume of Arctic sea ice reached its lowest ebb in the history of civilization during the fall of 2016 and has been declining ever since—it's also affecting the jet stream and the polar vortex that either traps cold blasts of air or lets them out. This has, Francis said, made the jet stream "wobbly" and uneven.

With temperatures soaring in the Arctic, the jet stream is getting even more wobbly. It's no longer the nice, smooth-flowing jet stream that meteorologists have grown accustomed to, which allows them to predict the weather. It bends in serpentine fashion now. It gets "stuck" in one pattern for days on end, wreaking havoc with local weather.

The biggest problem is that the jet stream wobble may be permanently changing nearly all weather patterns across the Northern Hemisphere . . . making everything unstable, unpredictable, and disruptive. Scientists may ultimately be able to prove that this new jet stream created a blocking ridge off the coast of California that essentially locked in years of drought, which has had a profound effect on the nation's agricultural breadbasket in the Central Valley.

The Arctic is perhaps the clearest example of the wolf at our door. It is trying to warn us that the extremely warm temperatures in the Arctic are now affecting the rest of the planet. The jet stream wobble is the manifestation of that warning. It isn't simply about polar bears. It's about the human species.

What happens in the Arctic doesn't stay in the Arctic. Big temperature increases there are permanently altering the region in ways that will have untold consequences. In fact, the Arctic system has changed so dramatically that it may now be vulnerable to tipping points that affect the entire planet.

For instance, a significant portion of the ice that covers the Arctic Ocean during the long winter used to be multiyear ice—which means that it was there over a long period of time and never really melted. That's no longer the case and partially explains why sea ice in the Arctic is dramatically lower than it's ever been at this time of year.

NASA satellites and scientists on the ground estimate that multiyear ice used to make up a fifth of the ice cover in the Arctic only thirty or forty years ago. By the end of 2016, it had shrunk to just 3 percent.[4]

The implications of this should be obvious. The planet has warmed rapidly in a very short period of time, over decades. And in the region most sensitive to that warming, the ice at the top of the world that used to remain there all the time is presently disappearing almost entirely during the summer. The multiyear ice that served as a barrier against ice disappearing entirely is gone. We will now see essentially ice-free summers in the Arctic sooner rather than later.

It is against this stark backdrop in the Arctic that an international group of scientists released a study in the winter of 2016—a study that laid out in painstaking detail why a series of imminent regime shifts could soon occur in the Arctic.[5]

The *Arctic Resilience Report* was five years in the making and is a first of its kind. It was published by the Stockholm Environment Institute under the direction of the Arctic Council—an official body that includes all of the nations that interact with the Arctic alongside groups representing indigenous populations there.

Russia is a member of the Arctic Council. So is the United States. Dozens and dozens of some of the world's best scientists who have devoted their professional lives to studying the Arctic contributed to the report. It is a sober, nuanced, objective, science-based effort to explain what might happen if tipping points are crossed in the Arctic thanks to global warming that is permanently altering the region.

The scientists didn't pull any punches in the report.

"The Arctic is now changing at an unprecedented pace, on multiple levels," wrote the study's cochairs, Johan Rockström and Joel Clement. "Arctic social and biophysical systems are deeply intertwined with our planet's social and biophysical systems, so rapid, dramatic and unexpected changes in this sensitive region are likely to be felt elsewhere. As we are often reminded, what happens in the Arctic doesn't stay in the Arctic."

The study first chronicles what scientists have known for some time. "Arctic ecosystems are changing in dramatic ways: ice is melting, sea levels are rising, coastal areas are eroding, permafrost is thawing, and landscapes are changing as the ranges of species shift," it said.

But we've known for some time now that the Arctic was being assaulted. That's just the opening act. It's what comes next that the report tried to address. The report was designed to warn people that they will need to adapt to the coming changes.

The report documents nineteen regime shifts in the Arctic that are under way, beyond just ice-free summers. These regime shifts are perhaps better known to the public as "tipping points"—the sorts of things that, once crossed, are irreversible. We can never go back once these regime shifts occur.

The report goes through each of them, ranging from the collapse of different Arctic fisheries to the complete transformation of landscapes like bogs, peatlands, tundra, or boreal forests. Once these regime shifts occur, it will profoundly affect much of the rest of the planet. The Arctic is the harbinger of what is to come. It is our canary.

Some of the tipping points seem innocuous—until we learn that changes in the Arctic will spread to the rest of the planet. When vegetation replaces ice, the region can no longer reflect heat away from the earth. Once methane is released, it accelerates other feedback loops. Shifts in snow distribution patterns could alter the monsoon season in Asia, where a billion people rely on its stability for their source of fresh water. A collapse in key Arctic fisheries could profoundly affect ocean ecosystems everywhere else.

"The consequences of some of these shifts are likely to be surprising and disruptive—particularly when multiple shifts occur at once," the report said. "By altering existing patterns of evaporation, heat transfer and winds, the impacts of Arctic regime shifts are likely to be transmitted to neighboring regions such as Europe, and impact the entire globe through physical, ecological and social connections."

The Arctic resilience report shows that these regime shifts to the climate and ecosystems are now just over the horizon. They will inevitably lead to geopolitical regime shifts, which national security experts have been warning us about for years. In fact, some have been preparing for precisely this eventuality.

A Department of Energy national laboratory that specializes in the intersection of national security threats and energy—Sandia National Laboratories—issued a special report in the spring of 2015 on precisely what the coming regime shifts in the Arctic mean for American national security interests.

"It is important to understand and predict future states of the Arctic for many reasons," the Sandia report said. "Accurate predictions based on numerical models for the future states of Earth systems are important for our energy future, for geopolitical reasons, and for national security. Although the Arctic contains a tiny fraction of the world's total population, its influence on the rest of the world, both in a geophysical and a geopolitical sense, is large."

Sandia identified a number of imminent threats in a changing Arctic, ranging from the fact that the region is considerably more sensitive to temperature extremes than other parts of Earth and that "rapid changes in the Arctic have a disproportionate effect on the surrounding northern continents. There is also considerable concern that Russian dominance of an ice-free Arctic could threaten European, Canadian and American interests.

"The Arctic has mineral wealth, water, fisheries, trade routes, and a militarily-strategic location," Sandia said. "Nations with Arctic borders depend on a natural defensive barrier against potential aggressors."

Despite the fact that it is essentially the control knob for Earth's entire ecosystem, it is one of the least understood parts of Earth—and a place where we are almost flying blind. "The Arctic thus cannot be analyzed in isolation. The Arctic is one of the least understood but most important climate subsystems in the Earth climate system," Sandia concluded in its report.[6]

That's why leading national security and foreign policy experts have called for much greater scrutiny of the intersection of military considerations and a rapidly changing climate in the Arctic region.

The Council on Foreign Relations was so concerned about the lack of attention being paid to the region that it convened a global task force in early 2017 to bring it directly to military and world leaders. The Alaskan Arctic is the "fourth coast of the United States," the CFR said, and it was time that military leaders recognized the imminent threats in the region.

"Things we see in the lower forty-eight [states] tend to converge in the Arctic in a way that should give us insights on future threats," said Thad Allen, who cochairs the CFR's Independent Task Force on the Arctic.[7]

It is time that the United States started to recognize that the rapid changes occurring in the Arctic have immense implications for U.S. strategic and economic interests, Allen argues. Right now, he said, the United States is being outflanked by Russia (which has the longest Arctic coastline and derives 95 percent of its natural gas and 75 percent of its oil from Arctic territories). But to date, American military leaders have been content to largely ignore the changes occurring in the Arctic.

"Alaska is critical to national security of the United States," Allen said. "It is where we have ground-based missile systems that can intercept potential missile launches from Asia. It has operating bases where aircraft can refuel. It is on the great circle route from the United States to Asia. If you look at the ability of U.S. forces in Alaska to deploy to the Pacific Rim, it becomes a very important forward operating base or stopover point."

Russia clearly understands this—and has moved aggressively as the region changes. "Russia has very aggressive plans for the Arctic in terms of resource extraction," Allen said. "It has made claims to an extended continental shelf, almost to the North Pole. There is no doubt that Russia has made a significant commitment both economically and politically to extend its presence, its access to resources, and to manage transportation routes close to the Russian coastline."

If the United States continues to ignore the rapid changes occurring in the Arctic region, it is essentially abandoning the defense of the northern part of its border, Allen argues.

"The Arctic represents the fourth coast of the United States," he said. "Its prominence, geographically, makes it an indispensable part of the national security structure of the United States."

At a minimum, the United States needs to have both a military and diplomatic presence in the region because changes are occurring so rapidly. "If we are to interact with Russia and deal with its more aggressive activities up there," he said, "we have to have a sustained presence."

7

The Sahel

The plan was simple, as was the story. The Sahara Desert was spreading throughout the region of North Africa known as the Sahel, the arid savanna that runs along the southern border of the most famous desert on the planet. More than a dozen countries that touched the desert in one way or another were all faced with the same terrible problem. The desert threatened to cripple their way of life.

Tens of thousands of farms were at risk of being swallowed by creeping desertification. The region strained under a crippling drought that had lasted far longer than anyone could remember. Spring was coming earlier, and rainfall was so uncertain that farmers simply didn't know what to expect any longer when they planted their crops for the coming year.

So analysts at the United Nations and governments throughout the region came up with a plan—just as Chinese leaders once had thought to hold back the Gobi Desert and other desert regions. They would plant a Great Green Wall of trees that would extend across all of northern Africa—3,360 miles long, 10 miles wide, right through the middle of a dozen countries from Senegal in the west to Djibouti in the east.

"The desert is a spreading cancer," said Abdoulaye Wade, who was Senegal's president from 2000 until 2012 and was the most prolific promoter of the Great Green Wall inside the halls of the United Nations and the boardrooms of international financiers like the World Bank. "We must fight it. That is why we have decided to join in this titanic battle."[1]

Leaders now talk about the imminent success of the Great Green Wall. About 15 percent of the imagined trees in a handful of countries have been planted. Once completed, it is meant to be the largest living structure on the planet. Leaders convinced the World Bank, the government of France, and others meeting at the Paris Climate Accord in December 2015 to pledge $4 billion to make the Great Green Wall a reality and to save the Sahel across all of northern Africa.

"The project has sky-high ambitions; to restore 50 million hectares of land, provide food security for 20 million people, create 350,000 jobs, and sequester 250 million tons of carbon," CNN said in a special report on the initiative.[2]

Here's the thing, though. The story is real—but it is also far too simple. And the solution everyone has rallied around may, in fact, be doomed to fail. An earlier effort to plant fifty million trees in Niger failed spectacularly. Almost 80 percent died within two months. "Of the 50 million seedlings planted every year in the 11 northern Nigeria states worst effected by desertification, 37.5 million wither and die within two months," a special 2008 UN report concluded.[3]

Just as China learned the hard way, simply planting trees to hold back the march of the desert—without any thought about all of the other things that make an ecosystem sustainable over the long haul—is certain to fail. Unless leaders change course and build new models into the Great Green Wall, the Sahel will fall much as Yemen, Syria, Somalia, and others have.

"This was a stupid way of restoring land in the Sahel," Dennis Garrity, a senior research fellow at the World Agroforestry Centre, told *Smithsonian* magazine for a special report on the Sahel (which includes Senegal, Mauritania, Mali, Burkina Faso, Algeria, Niger, Nigeria, Chad, Sudan, South Sudan, Eritrea, Cameroon, Central African Republic, and Ethiopia).[4]

"If all the trees that had been planted in the Sahara since the early 1980s had survived, it would look like Amazonia," Chris Reij, who works at the

World Resources Institute and has worked in Africa since the late 1970s, told the magazine. "Essentially 80 percent or more of planted trees have died."

Garrity, Reij, and other Sahel experts all say roughly the same thing—the Sahara is not really moving south. Rather, a combination of factors ranging from water scarcity, overuse, and drought is denuding the land, making it *appear* as if the Sahara is taking over parts of the countries on its southern border.

This, however, may be a distinction without a difference. Regardless of the combination of factors, the Sahel is in trouble. The desertification problem is a nearly unimaginable horror waiting to erupt soon. But its root causes are so complicated—and so intertwined in a combination of factors from drought and extreme rainfall to overuse and a population explosion in the region—that leaders have been forced to weave together a simple narrative with a simple solution.

What's facing the Sahel is just a wicked, wicked problem. There's no easy way to sugarcoat it. But scientists and researchers have argued for years about the real problems and the real solutions—the ones that will actually take root and transform the desert once again before it swallows all of North Africa whole.

The questions are huge. Is the Sahara Desert really moving south? Or are there other factors at play? When tens of thousands of farmers give up their farms throughout the countries in the region, is it because the desert has overwhelmed them? Or is it because extreme rainfall patterns, combined with crippling droughts that last longer than they used to, have made farming a shaky, uncertain enterprise?

Rainfall in the Sahel has always been low, averaging anywhere from four to twenty-four inches a year. Droughts have occurred on a regular basis. What has changed in recent years—making planting and harvesting nearly impossible to predict alongside the encroachment of the Sahara—is almost certainly connected to climate change. There is up to 8 percent more water vapor circulating in the atmosphere now, triggering intense and extreme rainfall patterns in places like the Sahel. Droughts now last much longer (sometimes for several years on end) than they have historically and will only get worse from here on out, climatologists say.

The science problems facing the Sahel are daunting in extreme ways. The easy solutions—like planting tens of millions of trees along the edge of the desert throughout the Sahel region to keep the sands at bay—is almost certainly a losing proposition and far too simplistic. The 2008 failure in Niger should have been a clear warning, but it wasn't. No one seems to have studied the collapse of a similar "green wall" initiative in China that failed to hold back the desert.

But one thing is certain. The Sahel is in trouble right now, and solutions need to find their way to the front of the line immediately. Various arms of the United Nations and risk analysis firms now say that one hundred million Africans are threatened by growing desertification in the Sahel.[5] That encroaching desert is driven, in part, by extreme rainfall patterns and the worst drought in thirty years across the continent . . . both of which are climate signals that scientists have studied at length.

Two-thirds of Africa's arable land could be lost in just the next ten years if that trend continues. This comes on top of the gravest humanitarian crisis—famine and starvation that could kill up to twenty million people in four countries at the Horn of Africa, first evidenced in the summer of 2017—in the history of the United Nations.

Without immediate solutions, young men who are forced to leave the land will turn to other, violent solutions or simply leave Africa as refugees to seek uncertain fortunes elsewhere, Elvis Paul Tangam, the African Union commissioner for the Sahara, told *Living on Earth* in 2016 as the refugee crisis washed over Europe.

"The Great Green Wall is about development; it's about sustainable, climate-smart development, at all levels," Tangam said. It's a matter of life or death for millions of Africans, especially young men.

Idle young men who have no work to turn to in the foreseeable future and have watched helplessly as crops fail and livestock die, Tangam said, are then faced with an impossible choice. They can join rebel or terrorist groups like Boko Haram or become part of a flood of refugees risking their lives to cross the Mediterranean to try to find work in Europe.

"Every young person wants to be valued. In the African context, every young person, especially a young man, has the responsibility to take care of their family," he said.

Virtually every story about migrants who have overwhelmed Europe's refugee system focuses on families that have left horrific military conflict in Syria and regions held by the Islamic State or fled from terrorist groups like Boko Haram. They focus obsessively on the worst aspects of the military conflict and what forces families to flee from the constant terror inflicted by the Islamic State and others.

They rarely talk about the underlying causes driving that conflict—and the uneasy thought that Africa is potentially on the cusp of losing two-thirds of its arable land in just ten years. That scenario is a recipe for revolution and terrorism that easily spreads far beyond North Africa. The Sahel is ground zero.

Countries in the Middle East and northern Africa—where water scarcity, the disappearance of arable land, poor land-use practices, and overpopulation have made it impossible to feed people—are now importing huge quantities of food. This has created a massive imbalance around the world.

An eye-opening new study in the spring of 2017 in the journal of the American Geophysical Union illustrates just how stark the situation has become—and why the fate of the Sahel will determine many futures in the next few years.[6]

Roughly *half* of the world's population now lives in areas where food imports are needed to make up for "food scarcity," meaning that more than three billion people live in areas where they rely on the government to make sure they have food to eat. One-fifth of the world currently depends entirely on food imports to survive.

While researchers have always known that countries in the Sahel and elsewhere were clearly struggling to feed their populations in the face of water scarcity and usable land constraints, no one had ever truly added up the magnitude of the problem facing half the world today. No one had ever linked trade—specifically, food imports—to deal with the situation. That's what the AGU study did.

The researchers in the AGU study created a simple metric to study the real, underlying problem. They compared net food imports in a region (e.g., the Sahel) to its ability to grow food. If a country is wealthy enough and can afford to purchase imported food, then a food import strategy works. In a significant number of instances, however, it won't.

While the study found that much of the Sahel and the Middle East are able to meet demands by importing food, that isn't the case in up to 15 percent of the world. That's a huge number of people living in places where the local agriculture can no longer feed the population and where they face starvation because the country isn't wealthy enough to import food. If things fall apart even further in these parts of the world, that number—15 percent—will grow exponentially.

"According to the sub-national estimates, food imports have nearly universally been used to overcome local limits to growth, affecting 3.0 billion people—81% of the population that is approaching or already exceeded local carrying capacity," the researchers wrote in summarizing the results of their study.

"This strategy is successful in 88% of the cases, being highly dependent on economic purchasing power. In the unsuccessful cases, increases in imports and local productivity have not kept pace with population growth, leaving 460 million people with insufficient food. Where the strategy has been successful, food security of 1.4 billion people has become dependent on imports."

The theory that there are defined limits that will keep the total population on Earth from growing beyond a certain point has been around for centuries. A scholar, Thomas Robert Malthus, proposed such limits in 1798. Later efforts, such as the Club of Rome's *The Limits to Growth* in 1972, likewise made the case that the human population would eventually reach its limit—that there would not be enough land available to feed everyone. Over time, technological innovation and a green revolution made it clear that big agribusinesses could, in fact, grow considerably more food than people had anticipated. The coming catastrophes imagined by Malthus and the Club of Rome never materialized. Agricultural innovation and the green revolution saved the day.

But what the hundreds of studies and theories assessing the limits of growth rarely took into account were the types of constraints confronting places like the Sahel today—a wicked combination of water scarcity that will only get worse, not better; the potential collapse of arable land; overuse as too many people crowd into the same locations; and an exploding population in the midst of all three.

Researchers are starting to recognize that a *combination* of factors lead to the existence of resource limits. But the scarcity of both land and water continue to take a backseat to discussions about population limits and environmental factors. That's what the novel AGU study set out to address.

"Current demographic theories rarely consider scarcity of resources such as water and land as limiting factors for population growth," the researchers wrote. "However, at local scale there are many areas where local limits to growth may already have been exceeded, as scarcity of water and land resources limit sufficient food production, despite the rapid technological and agronomic advancements increasing the efficiency of agriculture."

For this reason, in a place like the Sahel—where local limits to growth have almost already been exceeded—all that's left is the ability of the government and businesses to import food. That strategy works, but only as long as the governments have enough wealth to keep importing food, the AGU study concluded.

The policy implications of this study in the Sahel and elsewhere are enormous. Right now, even without further changes in the ecosystems of the Sahel and the Middle East, 1.4 billion people in these regions are now entirely dependent on food imports for their survival. The countries in the Sahel are essentially covering up their food security problems with food imports—a system that is perfectly suited for disruption by economic or military uncertainty.

"Many North African and Middle Eastern countries seem to be successful at importing food to meet demand, according to the study," AGU's

Lauren Lipuma wrote.[7] "Most of these countries are in arid regions where the environment can no longer support the local population, but since these countries are relatively wealthy, they have supplemented their limited food supply with imports."

That isn't the case in other parts of the world—and is a harbinger for the Sahel if the Great Green Wall fails. "India and Nepal . . . have been relatively unsuccessful at importing food to feed their growing populations despite increasing agricultural productivity and recent increases in economic power," Lipuma wrote. "Some of the more densely populated areas in India can't feed their local populations even with current levels of imports and food from other areas of the country."

There is some good news in the Sahel. Several countries in the region, especially Niger, appear to have learned from their mistakes. After the self-evident collapse of tree-planting in Niger almost a decade ago, farmers took matters into their own hands and literally went back to their roots. They found a cheap, effective way to bring "green" back to the Sahel—and to make it stick. Tens of thousands of farmers began to protect trees that grew up alongside their crops.

It worked. Green swaths returned to Niger, Burkina Faso, and Mali. It was a practice that had once been used in the late twentieth century—largely because farmers had seen with their own eyes that it kept the desert at bay. The combination of trees and crops kept the topsoil from blowing away and protected the land. Meanwhile, farmers in Burkina Faso built stone barriers around fields to keep water from running off, while simultaneously encouraging rainfall that occurred even in intense bursts from simply washing away. They also built pits near trees to store water. These two relatively simple indigenous land-use techniques—letting trees grow alongside crops, and carefully preserving water that goes to the roots of trees and crops—has seemingly worked when all else failed.

To its credit, leaders at the UN's Great Green Wall initiative have taken notice. They now tout the indigenous land-use techniques—the ones that tens of thousands of farmers in Niger and Burkina Faso had implemented without anyone telling them to do so—as the new heartbeat of the initiative. The African Union and the UN's Food and Agriculture Organization (FAO) now refer to it as "Africa's flagship initiative to combat degradation, desertification and drought."

"The [Great Green Wall] program has moved from forestry to sustainable land and water management," Tangam told CNN. "The ambition remains the same, but the activities have changed."[8]

The rest of the world may continue to think about it as the Great Green Wall of trees at the edge of the Sahara, but those closest to the initiative now know better. The big groups that have pledged funding seem to have caught the wave created under the radar by local farmers going back to their indigenous roots as well.

"We moved the vision of the Great Green Wall from one that was impractical to one that was practical," said Mohamed Bakarr, the lead environmental specialist for Global Environment Facility, which serves as an analyst for the World Bank on projects like the Great Green Wall. "It is not necessarily a physical wall, but rather a mosaic of land use practices that ultimately will meet the expectations of a wall. It has been transformed into a metaphorical thing."[9]

All of the globe-straddling financial, humanitarian, philanthropic, and conservation groups now glowingly tout the new strategic concept that seems to hold such promise for the Great Green Wall—a process that they learned from the local farmers in Niger and Burkina Faso who were simply doing what they knew would work because they had done it before. They call it "farmer-managed natural regeneration." Quite a few research studies are under way to track the phenomenon, especially as countries like Malawi and Ethiopia have heard about the practice and begun to adopt it locally.

Reij, the WRI specialist who has seen the progress in Niger, Burkina Faso, and Mali firsthand, is hopeful that the big NGO, humanitarian, and conservation groups will stand clear and simply encourage the farmers to do what they now know works. "If you want to re-green, do it quickly and effectively and at a reasonable cost, the only way forward is natural regeneration on farms," he said. "Put responsibility in the hands of the farmers. They know what their best interests are. Conventional projects will not make a difference here."

Reij said he and several other analysts almost overlooked the transformation that miraculously happened on the ground in Niger, Burkina Faso, and Mali. Satellites that routinely monitor the Sahara from space couldn't separate crops from trees. It was only when he was in an airplane flying over parts of these countries that the analysts discovered what the farmers had done. After glimpsing trees interspersed with the crops and entire areas turning green again, the analysts went to the local farms and asked farmers what they'd done. It was a stunning and humbling revelation.

In Niger, in the wake of the tree-planting fiasco, the local farmers had to ask the government for permission to grow trees the right way—the way that made the most sense for their own farms and for efforts to keep topsoil from blowing away. Thankfully, the government gave its blessing in Niger. That isn't always the case when governments and their big global donor partners take a top-down approach to problems. They routinely ignore what local experts say and do, opting instead for their own solutions (like spending $4 billion to plant a row of trees 3,360 miles long at the edge of the Sahara Desert).

The epiphanies can't come a moment too soon. The jury is still out whether a program that's centered around indigenous land-use techniques—rather than foolishly planting a row of trees 3,360 miles long at the edge of the Sahara—will save the Sahel and keep tens (or hundreds) of thousands of farmers from simply giving up in the face of a monstrous, overwhelming, unfolding catastrophe.

Each country along the Sahel is free to define the Great Green Wall as it sees fit. Some almost certainly will simply plant trees at the edge of the Sahara, only to see them die within a matter of months. Analysts familiar with the Sahel countries expect that many of them will never allow local farmers to lead the way with cheap, indigenous land-use techniques—not when these techniques aren't designed to attract huge donor investments from global philanthropies.

But just as nomads once found a way to spread the word quietly about the location of oases in the Sahara that sustained life at a time when maps did not exist, it's entirely possible that the farmers all along the Sahel at the southern edge of the desert, who genuinely know and care for the land, have found a way forward when one did not seem to exist even a few years ago.

8

Ocean Colonies

The Great Barrier Reef is the largest living structure on Earth. It's so massive that satellites and astronauts can observe it quite easily from space. The immense coral reef system is more than 1,400 miles long and runs along the northeastern coast of Australia.[1] It is one of the world's natural wonders.

It's also dying, before our eyes.

A peer-reviewed study in *Nature* magazine in early March 2017 surprised even the most cynical and jaded marine researchers who have been warning us for almost twenty years that ocean reefs were in serious trouble all over the world.[2]

The study, which was based on aerial views of the entire reef and then compared to historical records of earlier bleaching events in 1998 and 2002, found that much of the Great Barrier Reef in its northern section was now dead. Only a cyclone that cooled the waters at the southern end of the reef likely saved half of it.

The collapse and death of the Great Barrier Reef will have an enormous impact on the Australian economy. Tourism on the Great Barrier Reef employs seventy thousand people and generates $5 billion in income each year.[3]

Earth's three hottest years in the historical record were in 2014, 2015, and then 2016. Warmer ocean water during those three years put a great deal of stress on the great reef. But it wasn't until researchers actually looked at the images and maps up and down the coastline that they were able to gauge the true extent of the damage.

The *Nature* study was so definitive in its findings about the massive coral die-off, and its implications, that it likely ended whatever hope marine scientists might have harbored that potentially catastrophic impacts were still off in the distance somewhere. Scientists have been optimistic that the worst impacts might not occur until the end of the twenty-first century. That isn't the case any longer.

"Since the 1980s, rising sea surface temperatures . . . have triggered unprecedented mass bleaching of corals, including three pan-tropical events in 1998, 2010 and 2015/2016," the *Nature* researchers stated.[4]

It was the third global-scale mass bleaching event since marine scientists began studying the phenomenon beginning in the 1980s. Sea surface temperatures were responsible for all three global-scale events, they concluded after comparing "distinctive geographic footprints of recurrent bleaching" on the Great Barrier Reef.

Only "immediate global action to curb future warming (can) secure a future for coral reefs," the researchers concluded in the *Nature* study.

The lead researcher on the study, Terry Hughes from James Cook University in Australia, told *The New York Times* that his graduate students cried as he showed them maps of the damage that warming ocean waters had done to the great reef.[5]

Coral reefs—vast and complicated living structures that are the hub of an enormous web of ocean life—can turn white from bleaching if the water is too warm in the ocean. This bleaching, in turn, can cause portions of a reef to die-off for good.

That's what is happening to the Great Barrier Reef and to other coral reef systems all over the world in the face of unremitting carbon emissions that are causing the planet to warm.

"Corals are unable to cope with today's prolonged peaks in temperatures— they simply haven't been able to adapt to the higher base temperatures of the ocean," the Global Coral Bleaching consortium concluded recently.[6]

Coral reefs are formed when coral skeletons accumulate over hundreds or thousands of years. These skeletons, made of calcium carbonate, create a complex structure that supports a huge diversity of marine life.[7]

Coral die-offs can occur for several reasons. Water pollution or disease can irritate coral skeletons, which in turn causes them to expel the brightly colored algae that live within the coral systems in a symbiotic relationship. Both of these are complicated when a coral reef is stressed by warm water.

In fact, it's the warming ocean water—now occurring on a long-term basis from climate change and a short-term basis from weather events like El Niño—that is the primary reason brightly colored algae are expelled. That is what turns the coral reef white, known as "bleaching." Once the algae are gone, the reefs essentially "die" as the vast diversity of life goes away.[8]

The oceans, meanwhile, absorb even more of the carbon dioxide (created when we burn fossil fuels) than the atmosphere, which causes other problems for ocean life and will accelerate planetary warming in coming years.

For this reason, oceans may be an even earlier warning system that we're in serious trouble than rising temperatures on land. That's why dying coral reefs all over the world are a very clear signal that the changing climate system is starting to have profound impacts.

Until the March 2017 study in *Nature*, most scientists had assumed it would be thirty or forty years before warm ocean waters would begin to truly kill off the great coral reef systems like Australia's Great Barrier Reef.

Not now. The *Nature* study was an immense wake-up call for the science community.[9]

"We didn't expect to see this level of destruction to the Great Barrier Reef for another 30 years," Hughes said. "In the north, I saw hundreds of reefs—literally two-thirds of the reefs were dying and are now dead."[10]

Coral reefs are more than just colorful attractions that support a big tourism economy (though they are certainly that, especially in countries like Australia). If most of the world's coral reefs die—which now appears to be the case, sooner than scientists had expected—it will have an immense and immediate impact along coastlines across the planet. Florida, for instance, benefits from tourism associated with diving in coral reefs. The coral reefs off the Florida Keys help generate more than $1.6 billion in revenues annually.[11]

Hundreds of millions of people in many countries get their daily protein primarily from fish that live in and around coral reefs. If the reefs die, so do the fish that feed those populations. That turn of events would spark an unprecedented humanitarian crisis.

Researchers all over the world are finding coral reef systems that are experiencing die-offs now at an unprecedented rate. The Great Barrier Reef die-off is the most well known, the largest, and the most shocking—but others in various parts of the world are just as stressed.

The reefs might recover in fifteen or twenty years, but they might not. Their fate is inextricably linked to the health of the planet. If the planet continues to warm, there is little hope that these coral reefs can rebuild. A 2016 study in *Science* magazine concluded that the Great Barrier Reef may never recover.[12]

While coral reefs make up a very small portion (about 1 percent) of the marine environment in the world, roughly a quarter of all marine life makes their home in and around reefs. It's why so many fishing communities subsist off fish from coral reefs. When the reefs die, the fish do, too . . . and hundreds of millions of people lose their ability to feed their families from the marine life on the reefs. The death of coral also represents a huge loss—as much as $375 billion annually—for the local economies they support.[13]

New research from the University of Exeter in 2017 found that increased surface ocean temperatures led to a coral die-off in the Maldives, essentially causing reef growth rates there to collapse. "Similar magnitudes of coral death have been reported on many other reefs in the region . . . suggesting similar impacts may be very widespread," the university said in a release on the study.[14]

Photographer Richard Vevers quit his job to travel the planet to photograph the process of dying coral reefs that has now occurred on a continuing basis since 2014—the longest coral bleaching event in recorded human history. He founded a photographic collaboration with the University of

Queensland and other research institutions in order to chronicle the coral die-off around the world.

Like others, Vevers has chronicled the collapse of the Great Barrier Reef. "The soft corals were just decomposing—animals literally dripping off the rocks," he said. "The most horrifying part was that we just absolutely stank of rotting animals. That's when you really realize that reefs are made up of billions of animals."[15]

Vevers started his journey after the ongoing, worldwide coral bleaching event (again, the worst ever in recorded history) began in 2014. He wanted to capture as much of it as he could.

Coral reefs from Florida to Australia are dying, according to a report from the National Oceanic and Atmospheric Administration (NOAA). "There has been continual bleaching in the Pacific," says Mark Eakin, an NOAA coral reef scientist. "It's unlike anything we've ever seen before."

The die-off at the New Caledonia Barrier Reef Vevers photographed last year is what truly shocked him—even more than what he'd seen at the Great Barrier Reef in Australia. "I was blown away," Vevers told *Time* magazine. "I've never seen something so beautiful, but it's dying."

While scientists studying coral reefs are hopeful that the planet will start to cool and that the oceans will heal, they don't really hold out much hope.

"You can't grow back a 500-year old coral in 15 years," says Eakin. "In many cases, it's like you've killed the giant redwoods."

Other scientists are equally pessimistic. "If you think of corals as canaries [in a coal mine], they're chirping really loudly right now," Jennifer Koss, who runs NOAA's coral reef preservation program, said at a press conference on reef studies. "The ones that are still alive, that is."

Coral reefs aren't just dying off the coast of Australia. The same thing is happening off the southern coasts of America as well.

In 2016, sport divers found a similar mass die-off in the Gulf of Mexico one hundred miles offshore of Texas and Louisiana. The divers had traveled to the spot because it has historically been one of the healthiest coral regions in the world, and a place where sport diving is world-renowned.

What they found instead stunned them: "green, hazy water, huge patches of ugly white mats coating corals and sponges, and dead animals littering the bottom on the East Flower Garden Bank, a reef normally filled with color and marine life," NOAA's National Marine Sanctuaries' program reported. "The reef, which is part of Flower Garden Banks National Marine Sanctuary, is considered one of the healthiest anywhere in the region."[16]

The charter boat captain alerted marine scientists who had been monitoring the region for just such a die-off event. "The scientists are now reporting that a large-scale mortality event of unknown cause is under way on this bank," the NOAA report said. "The divers and researchers found unprecedented numbers of dying corals, sponges, sea urchins, brittle stars, clams and other invertebrates on large but separate patches of the reef."

The marine scientists subsequently found that up to 50 percent of the coral reefs off the coast of Texas and Louisiana—again, a region of coral reefs known worldwide as one of the healthiest and most colorful—were dead. Other charter boats started to find similar die-offs, NOAA said.

What this means is that coral reef die-offs are likely happening everywhere, in every ocean, right now. There are only so many marine scientists available to study coral reefs, and their budgets are being cut. They can only visit a small fraction of the reefs to monitor their health.

NASA earth science satellites, which President Trump wants to defund, can monitor some of the large ones like the Great Barrier Reef from space, but they can't assess the thousands of reefs that support the fishing subsistence livelihoods for millions of people around the world in local settings. We can only imagine what might be happening in each of these settings.

The world has lost half its coral reefs in the last thirty years, the Associated Press's Elena Becatoros reported earlier this year. "Scientists are now scrambling to ensure that at least a fraction of these unique ecosystems survives beyond the next three decades," she wrote in a report on coral die-offs in the Maldives. "The health of the planet depends on it: Coral reefs support a quarter of all marine species, as well as half a billion people around the world."[17]

While marine scientists had suspected that coral reefs were starting to buckle under the strain of the warming waters, they still felt that there was still time to see a recovery. Studies in 2016 and 2017 essentially ended that hope.

"This isn't something that's going to happen 100 years from now. We're losing them right now," University of Victoria marine biologist Julia Baum told AP. "We're losing them really quickly, much more quickly than I think any of us ever could have imagined."

Becatoros concluded on an especially ominous note. "Even if the world could halt global warming now, scientists still expect that more than 90 percent of corals will die by 2050," she said. "Without drastic intervention, we risk losing them all."

As if this isn't bad enough, local "dead zones" in the world's oceans are wreaking havoc on coral reefs as well. Scientists from the *Smithsonian* reported in the spring of 2017 in one of the largest peer-reviewed journals in the world—*Proceedings of the National Academy of Sciences of the United States of America*—that dead zones worldwide were now threatening reefs everywhere.

"Dead zones affect dozens of coral reefs around the world and threaten hundreds more," the *Smithsonian* reported on the PNAS study. "Watching a massive coral reef die-off on the Caribbean coast of Panama, they suspected it was caused by a dead zone—a low-oxygen area that snuffs out marine life."[18]

After years of wondering when impacts would start to occur on a mass scale, scientists and journalists alike are beginning to understand that it was a huge mistake to believe large-scale impacts wouldn't start to happen until the end of the twenty-first century when it would be someone else's problem.

The coral reef die-offs all over the world—which became evident and obvious to marine scientists, sport divers, and nature photographers alike—have jolted chroniclers out of a collective, somnambulant state of wishful thinking that potentially catastrophic events might be well off in future decades.

"The measured warming of the planet is not hypothetical. Nor are its effects, which are happening now, not decades from now," the *Washington Post*'s editorial board wrote in March 2017 in a blunt editorial.[19]

"An ecological catastrophe is unfolding off Australia's coast: Humans are killing the Great Barrier Reef, one of the world's greatest natural wonders, and there's nothing Australians on their own can do about it. We are all responsible," it said.

The Washington Post said that the Great Barrier Reef die-off study in *Nature* ended whatever doubts anyone harbored that other factors (like local pollution) might somehow be responsible.

"There is little doubt that temperature is the culprit," the *Post*'s editorial board concluded. "Reefs far away from human runoff and other local risks are suffering. Corals in pristine water bleached just like those in dirty water. The *Nature* study quantified a relationship between exposure to warm water and the severity of observed bleaching."

The Washington Post was especially critical of the role that Republican Party leaders in the United States are now playing in the face of such large-scale impacts and events. While a majority of GOP voters now understand that carbon emissions are causing harm and would like to see the American government do something about it, GOP political leaders have chosen to avoid or ignore the obvious for years.

"The Trump administration proposed deep cuts for the Environmental Protection Agency, singling out climate programs, as well as the National Oceanic and Atmospheric Administration, which monitors Earth's seas and skies," the *Post* said. "President Trump also began what will no doubt be a broad rollback of Obama administration climate rules."

Trump's decision to withdraw the U.S. from the Paris Climate Accord several months later, in June 2017, signaled to the world that climate change wasn't a priority for the country's leaders. Though he has since said he'd reconsider that position, he has also supported the oil industry and dismissed the importance of renewable energy.

"In the long run, the planet will change enough—hurting enough people in the process—that even Republicans will have to admit the issue must be addressed," the *Post* article stated. "The question is what price the nation and the world will pay, in dollars, lives and ecological catastrophe, because our leaders were negligent in the meantime."

While Americans may not be affected—yet—by the coral die-offs, hundreds of millions of people in fishing communities in other parts of the world are presently in jeopardy, a report by a consortium of researchers and civil society groups concluded.

"More than 500 million people worldwide depend on [coral reefs] for food, storm protection, jobs, and recreation," said the International Union for Conservation of *Nature*. The coral die-offs threaten this entire population.[20]

Coral reefs are an important part of every local community along ocean coastlines. A third of the world's coral reefs—and a quarter of all fish species—are found off the coast of Indonesia. More than 90 percent of Indonesia's coral reefs are in trouble.[21]

In the Philippines, only 5 percent of the coral reefs are in "excellent condition." Coral reefs off the coast of Taiwan have gone beyond white bleaching and are literally turning black from disease.

More than 50 percent of coral cover in the Caribbean has now disappeared. A recent United Nations report said that most of the coral reef cover in the Caribbean will simply disappear in the next twenty years.[22] A combination of warm water, overfishing, and local pollution has combined to signal the extinction of reefs there.

"[A] population explosion along the coast lines, overfishing, the pollution of coastal areas, global warming and invasive species . . . have put Caribbean coral reefs in danger of extinction," the UN Environment Program concluded.

The UN report two years ago was exhaustive. It was based on an analysis of thirty-five thousand surveys taken from ninety different locations in the Caribbean. The study concluded that the corals that have suffered the most tragic declines are those in Jamaica, along the shores of Florida, and the Virgin Islands.

While most of us will never dive in the ocean to visit or photograph the lively, colorful coral reefs that have grown up along our coastlines over thousands of years, their extinction will nevertheless profoundly affect all of us.

Reefs provide a home for many species of fish that feed tens of millions of people. They support tens of billions of tourist dollars. Their collapse will affect those on the frontline of this extinction first—and the rest of us soon after.

That's why scientists and other experts are taking a hard look at the complex, insidious effects of climate change. With regime shifts clearly happening in places such as the Arctic, the Sahel, and the Great Barrier Reef, the time to act is now, before the issues become irreversible tipping points.

PART 3

The Impacts

Devastating changes are occurring right now across the globe, altering species, cities, farming, and other critical aspects of life as we know it.

Extinction-level dangers now face multiple species, including humans and vanishing iconic animal species, as a result of climate change. "The Impacts" tells the powerful stories of four of these devastating changes.

More frequent heat waves now threaten the lives of people across the planet—whether in the Midwest United States, Western Europe, Russia, or Bandar Mahshahr. As the planet warms, civil unrest and violence are spreading, especially in areas such as the Middle East.

Fueled by record-breaking ocean temperatures, rising sea levels, and precipitation changes, record storms are growing in frequency and intensity so much that a new category of superstorms might need to be created. These "Category 6" storms (with winds higher than 175 mph) have the capability of taking out coastal cities—even large ones such as Dubai and Tampa—and destroying the livelihood of fishermen and farmers.

Extreme events—whether storms battering coastlines, floods, or water scarcity—are creating a new but yet unrecognized category of refugees. These displaced people are a growing global concern for humanitarians and leaders of the countries where they migrate.

Iconic species such as the saiga antelope in central Kazakhstan have always been threatened by poaching, land development, and predatory wolves. Still, they survived for millennia—until a mass die-off of nearly two hundred thousand happened overnight. The African elephant and the giant panda in China also face extinction-level threats—largely because of changes in the environmental landscape that they may not be equipped to handle.

No doubt, the growing impacts in these four areas will alter our planet and its inhabitants, but if we prepare now locally, nationally, and globally with sustainable strategies, we can minimize the damage of heat, storms, and flooding and preserve the remnants of iconic species.

9

Dome of Heat

In the summer of 2015, at the northern edge of the Persian Gulf in Iran, the city of Bandar Mahshahr set an unofficial national record. The Iranian city hit a heat index of 165 degrees Fahrenheit. It became the second-highest heat index in history. Dhahran, Saudi Arabia, had reached a heat index of 178 degrees a few years earlier.

The Bandar Mahshahr extreme heat index came in the middle of a punishing heat wave that waxed and waned throughout the summer in the Middle East. Other countries nearby saw similar extreme heat index temperatures.[1]

Trying to function in that sort of temperature is like being forced to live and work inside a sauna that's cranked up high with the doors sealed. If temperatures at that level last for days, humans can't survive. It's a killing heat wave.

There are no official records for heat index temperatures, which combine heat and humidity. A heat index is actually a more relevant public health measure of temperature because it takes our ability to cool off in extreme heat into account. A "wet-bulb" temperature (as opposed to a "dry-bulb" temperature) takes other factors beyond just heat into account. Humidity is the most important factor.

The term itself, "wet bulb," sounds innocuous—a bit fuzzy, and even a little funny. It's anything but that. When heat wave temperatures in urban cities in America hit wet-bulb conditions, hundreds of people in the past have died in the middle of such heat waves that short out the power grid and cause suffocating conditions for populations that can't cope with them.

A wet-bulb temperature is essentially the combination of heat and humidity. Together they can be lethal, even if the heat doesn't seem quite so extreme. A wet-bulb temperature is measured with a thermometer wrapped in a wet cloth, distinguishing it from the commonly reported dry-bulb temperature, measured in open air. It is a measure of how well our skin can be cooled by sweating, which is how humans stay alive in the worst heat. But high humidity can defeat that cooling system; it makes the heat that much more dangerous.

Such wet-bulb heat waves don't only affect tropical, developing countries; they're a threat throughout the world. The July 1995 heat wave in the Midwest killed seven hundred. The 2003 heat wave in Western Europe killed forty-five thousand people. The 2010 heat wave in western Russia killed fifty-four thousand people.

Such wet-bulb temperatures are rare, but they might not be much longer. We're already starting to see the deadly combinations (like the 2010 heat wave in Russia) on a more regular basis. Scientists say that not only are we starting to see such dangerous wet-bulb conditions occur more regularly, they're also getting worse. In fact, we're now four times as likely to see such wet-bulb conditions as we were before the planet started changing. The sobering truth is that we used to see perhaps one wet-bulb day a year in America. Soon, we'll start seeing twenty or thirty of those kinds of days each year, and tens of millions of lives will be at risk.

A human's core temperature is about 98.6 degrees, but the skin temperature of the trunk is 4–9 degrees colder, depending on how warm it is and how active a person is. But sweating, which helps keep the core body temperature constant, becomes increasingly ineffective in increasingly humid air, and it can never cool the skin to below the wet-bulb temperature.

A person who is physically active at a wet-bulb temperature of 80 degrees will have trouble maintaining a constant core temperature and risks overheating. A sedentary person in the shade will run into the same problem at a wet-bulb temperature of 92 degrees. A wet-bulb temperature of 95 degrees is lethal after about six hours. Without air-conditioning, there's no escape.

It's hard to understate the implications of what wet-bulb temperatures hitting many parts of the world twenty or thirty days a year will mean. Entire cities have descended into nearly unmanageable chaos during one day of such wet-bulb temperatures in the past. Riots erupted over a single day of such unlivable conditions. What happens when these wet-bulb days descend on congested urban areas ten or twenty times as often? Unfortunately, we may soon begin to find out.[2]

That's why the temperature in Bandar Mahshahr in the summer of 2015 was potentially so deadly. At a heat index measurement of 165 degrees Fahrenheit, there is no real way for any human being to cool off. It's simply too hot for human life (or nearly any other life) over an extended period of time. If temperatures were to remain that high for days on end, those without the means to keep cool for vast parts of the day face life-threatening conditions.

When extreme heat waves last for days—which is now occurring more frequently in the subtropics, which includes the Middle East and parts of Africa—there isn't much that human beings can do while it's occurring. Police and military also go on heightened alert during heat waves, because history has shown that violence erupts much more quickly at such times.

Here's the warning that the Weather Channel issued during that Middle Eastern heat wave in the summer of 2015:

> The government has urged residents to stay out of the sun and drink plenty of water, but for many of the more than 3 million Iraqis displaced by violent conflict, that poses a dilemma. Chronic electricity and water cuts in Iraq and other conflict-ridden countries make heat waves . . . unbearable—particularly for the more than 14 million people displaced by violence across the region.
>
> In the southern Iraqi city of Basrah earlier this month, protesters

clashed with police as they demonstrated for better power services, leaving one person dead. Unlike other countries in the region, Iraq lacks beaches and travel restrictions make it difficult for people to escape the sweltering heat, leaving many—even those fortunate enough to live in their homes—with limited options for cooling off. Some swim in rivers and irrigation canals, while others spend these days in air-conditioned shopping malls.[3]

Just as water riots in Yemen were clear, unmistakable predictors of a violent, brutal civil war that would erupt two years later, and just as crippling droughts and extreme rainfall patterns in Syria forced tens of thousands of farmers to flee to big cities where there were no jobs, leading to civil unrest there as well, so, too, are these killing heat waves early warning flares of what's about to happen in the Middle East.

The planet is now growing warmer almost every single year. The ten hottest years in recorded human history have occurred since 2000. There are fluctuations occasionally in years where the El Niño/La Niña cycle has a big influence. The trend, the only scientific story that matters, is going in one direction—up.

But it isn't the gradual and incremental increases in global temperatures that worry public health and military leaders. It's the fact that extreme events like killing heat waves with wet-bulb temperatures reaching absurd levels (precisely what we saw in the Middle East in the summer of 2015) are becoming much more common.

During the twentieth century, less than 1 percent of the surface of the earth regularly experienced "extremely hot" events like what we saw in Bandar Mahshahr, NASA scientists reported in 2012.[4] But since 2006, about 10 percent of land area across the Northern Hemisphere has experienced these temperatures each summer, NASA said. That twentyfold increase is a huge change in such a short period of time.

"Global maps of temperature anomalies show that heat waves in Texas, Oklahoma and Mexico in 2011, and in the Middle East, Western Asia and Eastern Europe in 2010 fall into the new 'extremely hot' category," NASA said of the 2012 study published in *Proceedings of the National Academy of Sciences of the United States of America.*

What this means is that "extremely hot" events are becoming much more common and routine and are affecting significant numbers of people across the planet. They may be much more extreme in places like the Middle East, but these events will very quickly affect people outside the subtropics.

Another gauge of what's happening is the fact that record-high temperatures are routinely set in cities all over the world now on a consistent basis—and that record-cold temperatures occur half as much. Record-breaking high temperatures outnumber record lows by an average ratio of 2:1 on a decade-by-decade basis just in the United States. That ratio is even more out of balance in the rest of the world. In a normal world, where temperatures aren't rising, that ratio should be 1:1.[5]

Even "dry" temperatures (a measurement only of straight heat, with nothing else factored in) are setting mind-numbing records in some cities on a regular basis. Dry temperatures are what nearly all of us are used to hearing about. It's what official records measure.

Mitribah, Kuwait, set the official temperature as the hottest day ever recorded during the summer of 2016. The second-hottest temperature in history was recorded the next day in Basra, Iraq. It was 129.2 degrees Fahrenheit in Mitribah on July 21, 2016. It hit 129.0 degrees the next day in Basra.

There's some competition for the record. Death Valley in California—which is what many people in the United States associate with suffocating summer heat—has claimed various records that compare to what the people of Mitribah and Basra saw in the summer of 2016. Death Valley likewise hit 129 degrees in June that summer.

"While the Middle East's highest temperatures have occurred in arid, land-locked locations, locations along the much more sultry Persian Gulf and Gulf of Oman have faced the most oppressive combination of heat and humidity," meteorologist Jason Samenow wrote in *The Washington Post.* "Air temperatures of about 100 degrees (38 Celsius) combined with astronomical humidity levels have pushed heat index values, which reflect how hot the air feels, literally off the charts."[6]

The truth now, Samenow wrote, is that extremely high temperatures combined with humidity are hitting levels so high in the Middle East that we don't even have an index high enough to measure it.

"[This] combination of temperature and humidity is so extreme that it's beyond levels the heat index is designed to measure," he said. "The index, developed by R. G. Steadman in 1979, is actually only intended to compute values up to about 136 degrees." The heat indexes in Kuwait, Iraq, Iran, and elsewhere in 2015 were thirty degrees *higher* than the upper limit of what experts have used since the 1970s.

While it should be a warning, people outside the Middle East may not yet realize what it means. The record hot temperatures will become normal there and will spread to other parts of the earth.

"The torrid conditions observed in the Middle East over the last two summers may be a harbinger of even more extreme heat in the future," Samenow concluded for readers who follow his weather blog at the *Post*. "A study published in . . . October cautioned that by the end of the century, due to climate change, temperatures may become too hot for human survival."

It's already too hot for human survival in some parts of the world. The interior of Australia gets so hot now during the summer that weather forecasters and meteorologists had to come up with a new color code for weather maps for the interior of the continent. Forecast temperatures for Australia were so extreme that the weather bureau added a new color—purple—to its temperature scale, increasing the previous cap of 120 degrees Fahrenheit to 129 degrees.[7]

Australia is "turning the volume of extreme weather up, Spinal-Tap-style, to 11," the *Guardian*'s environment editor, Damian Carrington, wrote at the time. "The temperature forecast for next Monday by Australia's Bureau of Meteorology is so unprecedented—over 52C (125F)—that it has had to add a new colour to the top of its scale, a suitably incandescent purple."[8]

The reason was that the temperatures in the interior of Australia were much hotter than anything ever seen—and were too hot for the charts that they'd used since the 1960s. The thermometer eventually hit 54C in some places in the interior of Australia, shattering heat records by more than 3C.

"What makes this event quite exceptional is how widespread and intense it's been," Aaron Coutts-Smith, the weather bureau's climate services manager, told *The Guardian*. "We have been breaking records across all states and territories in Australia over the course of the event so far."

Besides making the temperature almost too hot for humans to survive in, the intense heat was triggering wildfires almost everywhere in the interior of the continent. Wildfires raged across New South Wales and Tasmania.

Scientists genuinely hate trying to measure how any one, single extreme event might (or might not) be connected to the overall change in Earth's climate system, but they have no choice now. "Attribution" studies, which are capable of explicitly tying a single extreme event like a deadly heat wave to the changing climate, are becoming much more common in the peer-reviewed scientific literature.

Researchers came together to produce an extraordinary set of studies that specifically looked at 2015 extreme weather events all over the world. All had a firm connection to Earth's changing climate, they said. Extreme heat was especially noticeable.

"As observed in years past, all the papers that looked at heat events around the world—from Egypt, Australia, Europe, Indonesia, Asia, India, and Pakistan—all found that climate change played a role in increasing the severity of the event," they wrote in the conclusion to the set of research papers published in *The Bulletin of the American Meteorological Society*.[9]

Scientists firmly linked the "dome of heat" rolling temperatures and heat waves in Australia to the changing climate.

Scientists have now had a chance to study the European heat wave of 2003, which killed forty-five thousand people, in much closer detail. Earth's changing climate system made that heat wave at least twice as likely.[10]

Likewise, a Russian heat wave in 2010—which killed fifty-four thousand people, wiped out $15 billion in wheat crops and directly led to the bread riots in Egypt's revolution—was three times as likely to have been caused by the overall change in Earth's system, scientists have concluded. That single event led to the warmest summer in Europe in the past five hundred years.[11]

All of this will only get worse from here on out. Mega–heat waves that kill tens of thousands—or, more likely, hundreds of thousands—of people will occur at least once every decade. They'll be up to ten times as likely to occur at any given time in the next few decades, researchers from Portugal, Germany, Switzerland, and Spain reported in *Science* magazine.

"Mega-heat waves such as the 2003 and 2010 events broke the 500-year-long seasonal temperature records over [half] of Europe," they wrote.[12]

None of this is especially surprising to people living in Australia. The interior of Australia has never been a fun place to live. Most of Australia's nineteen million people live near the coast because so much of the interior is desert, but in the past few years, Australia's interior has heated up in the summer to the point that it is at the edge of being too hot to support life. That's when news outlets added the new "purple zone" over the top of its previous meteorological charts in order to explain the much higher temperatures that had never been seen in central Australia.

"Severe and extreme heat waves have taken more lives than any other natural hazard in Australia," says writer Rae Johnston. "For example, during the 2009 Victorian bushfires, 173 people perished as a direct result of the fires; however 374 people lost their lives in the heat wave that occurred before the bushfires."[13]

But it grew exponentially worse starting in 2013, when news commentators first began to refer to a "dome of heat" so intense that it literally created roiling heat waves across the continent and raging wildfires. The weather bureau said portions of central Australia could see temperatures in excess of 120 degrees on a regular basis each summer. The all-time record high in Australia was 123 degrees set in 1960. New records have been set in every year since 2013 as the purple zone has become commonplace.

For now, these killing heat waves are occurring on a more regular basis in places where hardly anyone lives or where people are somewhat accustomed to going inside to get out of the heat when it becomes unbearable, but that won't always be the case. And, in some parts of the world, the extreme heat will become a tinderbox for unrest and conflict.

Basra in southern Iraq, for example, is one of the most coveted prizes for the Islamic State. Basra is stable—for now—but if it starts to become a regular part of the purple zone, it is merely a matter of time before this uncertainty adds a level of complexity that national security experts have anticipated for years.

The blistering heat and torrid conditions observed in the Middle East during these past two summers may be a harbinger of even more extreme heat in the future. Heat waves in India in 2015 killed 2,500 people. A heat wave in Pakistan that same year killed 2,000. At some point, it could simply be too hot to live in some parts of the world. A study published in the journal *Nature Climate Change* in 2015 used sophisticated supercomputer climate models to show that temperatures may become too hot for human survival at some point this century.[14]

Right now, the purple zone is confined to places like central Australia and Death Valley. The Middle East saw its own purple zone future for a few days at a stretch in recent summers. So did India and Pakistan. All of them were deadly. Extreme heat waves have killed tens of thousands of people at a time. They will grow more frequent until they reach a point where we come to expect them as commonplace.

This is our future. Some of it is happening right now.

"These heat waves will only become more common as the planet continues to warm. They don't just affect tropical, developing countries; they're a threat throughout the world," three earth sciences professors and researchers wrote in *The New York Times* in the aftermath of the deadly 2015 heat waves in India.[15]

Americans only experience dangerously humid days—wet-bulb high-temperature days where the body's core temperature may not be able to cool itself—just four days a year on average right now, wrote Robert Kopp, Matthew Huber, and Jonathan Buzan. Within a decade or so, they wrote, Americans will see ten such dangerously humid days—the type of days that can kill someone if they're not prepared.

People in America—the wealthiest nation in history—can afford to adapt to these killing wet-bulb days as they start to become commonplace soon.

"Since we can't avoid it now," the earth scientists wrote, "we must make our communities more resilient to heat and humidity extremes.

One step is to expand access to air-conditioning for those who can't afford it. We must also improve cooling in stiflingly hot factories and warehouses, strengthen public health systems, improve public warnings when heat and humidity are dangerously high, and be willing to shift outdoor work schedules."

That's fine for wealthy countries like America that can afford it, but for others, which won't be able to make air-conditioning a common good for people? "Some summers [will] have days so stiflingly muggy that a healthy individual [will] suffer heat stroke in less than an hour of moderate, shaded activity outside," they wrote.

Australia, and its new purple zone, is a cautionary tale for the rest of us. Even though Australians are used to extraordinarily hot weather, the past few years have been above and beyond what even they are accustomed to. Shops all across Sydney ran out of fans during the most recent heat waves. Political leaders encouraged people to go to the movie theaters to sit out the heat in the middle of the day. There were rolling blackouts from an overloaded power grid.

But Australia is also a bit fortunate. Because so many Australians live near the coast already, when the next batch of heat waves hit in 2017— marking the fifth year in a row that people were forced to deal with them— they were able to do what tens of millions of others around the world might not be able to manage when deadly heat waves become commonplace. They went swimming in the ocean.[16]

10

Category 6

Imagine for a moment that a tropical storm came along that was so powerful—with winds so strong that they were outside the normal range of what weather forecasters were accustomed to seeing—that it threatened to sweep aside almost everything in its path.

Now, imagine the level of devastation that such a tropical storm—called a "hurricane," a "typhoon," or a "cyclone," depending on where it happens— might inflict on people, buildings, and communities if its wind velocity managed to stay well above two hundred miles an hour when it makes landfall.

Finally, imagine more of these types of super-typhoons and mega-hurricanes occurring more often, while also forming earlier in the traditional hurricane season, reaching super-critical wind speeds later in the season, and occasionally developing weeks after the end of that season.

Sadly, there's no need to imagine it. A very rapid succession of four hurricanes (Harvey, Irma, Jose, and Maria)—all coming off extraordinarily warm water in the Atlantic Ocean, riding on rising sea levels that contribute to storm surges, and fueled by extra precipitation now regularly circulating in the atmosphere—swamped Houston, threatened all of southern Florida, destroyed entire islands in the Caribbean, and left Puerto Rico without power for months in the fall of 2017. These four hurricanes all reached incredibly high wind speeds over the Atlantic and only slowed down when they hit land. We've never seen such sustained wind velocities this late in the hurricane season in the Atlantic. We've also never seen the sort of extreme precipitation in such a short time span that flooded Houston.

Here's what we know about hurricanes like Harvey, Irma, and Maria. Record storms are fueled by record-breaking ocean temperatures, and rising seas and heavier rains trigger floods. Harvey, Irma, and Maria all intensified over above-average warm ocean waters. Global ocean temperatures were the warmest on record in 2016 and 2017. In the case of Irma, unusually warm waters in the Atlantic fueled the storm's record-breaking wind speeds. Irma maintained wind speeds of more than 180 mph for thirty-seven hours, longer than any storm—ever—on Earth in recorded history. As for Harvey and Maria, the warmer atmosphere loaded both storms with extra moisture, adding to the rain that kept Houston underwater for days, and led to floods and landslides in Puerto Rico. Maria was the third-strongest storm to make landfall in the United States, hitting Puerto Rico with 155 mph winds—just two miles per hour shy of Category 5 status.

Storms are becoming more difficult to predict and growing more intense. Extremely rare "five hundred–year" and "one thousand–year" events are occurring more often . . . even in back-to-back years. Ample research on Atlantic hurricanes shows that rising seas are dramatically extending the reach of storm surge in coastal communities. The dramatic flooding in Florida, Georgia, the Carolinas, and Virginia during Hurricane Matthew in 2016 is just one example of this trend.[1]

This very type of phenomenon is cropping up everywhere in the world. It only became real to America—for a brief moment in time—because four hurricanes threatened our entire southern coastline for a two-week stretch. But, until recently, we generally didn't hear the causal connections about super-typhoons or devastating hurricanes—at least, not yet—because scientists have been arguing among themselves for quite some time about how much hurricanes and typhoons are truly connected to the changing climate system.

Recently, though, the peer-reviewed science has taken a decidedly firm turn toward connecting hurricanes and typhoons to the changing Earth system. In short, hurricanes are forming earlier and later out over warm ocean waters and can now be stronger than they've ever been before.

The ingredients for massive tropical storms are well established: a pre-existing weather disturbance (e.g., a common storm), warm water in tropical oceans, moisture, and light winds. These elements are then combined into a big event—a hurricane or a typhoon. If this happens in the Atlantic or Northeast Pacific Oceans, it's called a *hurricane*. It's called a *typhoon* in the Northwest Pacific. It's a *cyclone* if it happens in the South Pacific or the Indian Ocean.

Once a tropical cyclone reaches maximum sustained winds of 74 miles an hour or higher, it's classified as a hurricane, typhoon, or cyclone. At that point, the weather event is likely to include violent winds, incredible waves, torrential rains, and huge floods. Weather forecasters rate hurricanes based on the strength of the winds. A Category 1 hurricane has slower wind speed. A Category 5 hurricane, which is rare, has winds above 155 miles an hour.[2]

There is no such thing as a Category 6 hurricane or tropical storm. The highest level—the top of the scale for the most powerful, most devastating hurricane or tropical storm capable of destroying entire cities like New Orleans or New York—is a Category 5 storm. Meteorologists and scientists never imagined that there would be a need for a Category 6 storm, with winds that exceed two hundred miles per hour on a sustained basis, sweeping away everything in its path. Until now, such a storm wasn't possible, so there was no need for a new category above Category 5.

Right now, however, there is anywhere from 5 to 8 percent more water vapor circulating throughout the atmosphere than there was a generation ago. This, combined with warmer temperatures that are driving water up from the deep ocean in places where hurricanes typically form, has created the potential for superstorms that we haven't seen before—and aren't really prepared for.

This combination of warmer oceans and more water in the earth's atmosphere—whipsawed by sustained periods of drier and wetter conditions in regions of the world that create superstorms—is now starting to create storms with conditions that look precisely what a Category 6 hurricane would look like. We might not *call* it a Category 6 hurricane for now, or know where to place it on the scale, but it doesn't mean that such previously unthinkable superstorms aren't about to start destroying parts of civilization in countries directly in their path.

The truth is that we've likely *already* seen a Category 6 hurricane in the past thirty years because evidence has mounted that hurricanes have become more extreme during that time period. The sheer number of hurricanes worldwide may have flattened in recent years, but warming has accelerated the strength of them in this era. Meteorologists have tracked several of them where winds were in excess of two hundred miles per hour at one point during their cycle over the oceans.

The Saffir-Simpson Hurricane Wind Scale is the scale that everyone uses when they describe the strength of hurricanes as they bear down on major cities along coastlines. Category 1 storms range from 74 to 95 miles per hour. The categories then go up in increments of 20 miles per hour. A Category 5 tropical superstorm (currently the highest we've ever needed) has sustained winds greater than 155 miles per hour. So if there is a Category 6 hurricane, it would be one with winds higher than 175 miles per hour. We're starting to see them now, experts say.

When Hurricane Katrina hit New Orleans, it began as a Category 5 hurricane in the Gulf. By the time it reached the southern coastline of the United States, it had fallen to Category 3. It still destroyed much of the city. Despite the fact that Category 4 and 5 storms have doubled in other parts of the world since 1970, only three Category 5 storms have reached America in the past century, including Camille in 1969 and Andrew in 1992. Several of the 2017 Atlantic hurricanes that hit the southern U.S. coastline were Category 5 storms over the ocean—and, in one case, stayed that high longer than others in history—but had slowed a bit before they reached the continental United States.

No one in America has ever experienced the wrath and fury of a Category 6 hurricane, which now genuinely seems possible and realistic. We've been lucky. Unofficial Category 6 hurricanes have appeared in other parts of the world, and we're seeing much stronger storms on a regular basis. It's only a matter of time before one hits America. When it does, it will come as quite a shock. The devastation we saw in 2017 in Houston, several Caribbean islands, and Puerto Rico may actually pale in comparison.

Jeff Masters, one of the most respected meteorologists in America, has begun to wonder publicly about the potential for a Category 6 hurricane. He launched a lively debate among his colleagues with a provocative post in July 2016 on the Weather Underground—a thought-provoking piece that prompted the Weather Channel and others to weigh in with their thoughts and theories as well.

"A 'black swan' hurricane—a storm so extreme and wholly unprecedented that no one could have expected it—hit the Lesser Antilles Islands in October 1780," Masters wrote to open the post. "Deservedly called The Great Hurricane of 1780, no Atlantic hurricane in history has matched its death toll of 22,000. So intense were the winds of the Great Hurricane that it peeled the bark off of trees—something only EF5 tornadoes with winds in excess of 200 mph have been known to do."[3]

Tornadoes on land have been known to reach wind speeds of up to 320 miles per hour. They can be quite deadly in local areas where they touch down. Hurricanes are much broader and can destroy entire cities under certain circumstances. After two devastating tornado events in 1997 and 1999, a new scale was devised to account for the most extreme tornadoes. EF5 tornadoes are the most violent of this type of meteorological event. There have been fifty-nine of these violent tornado events (either EF5 or the earlier F5 classifications) in the United States since 1950.[4]

Masters then made the startling claim that such a "black swan" hurricane was not only *possible* now but almost *certain* to occur more than once. He said that such storms should more properly be called "grey swan" hurricanes because the emerging science clearly showed that such "bark-stripping" mega-storms are nearly certain to start appearing.

"Hurricanes even more extreme than the Great Hurricane of 1780 can occur in a warming climate, and can be anticipated by combining physical knowledge with historical data," wrote Masters, who once flew into the strongest hurricane at the time as one of NOAA's "Hurricane Hunters" in the 1980s. "Such storms, which have never occurred in the historical record, can be referred to as 'grey swan' hurricanes."[5]

Masters based his bold prediction on research by two of the best hurricane scientists in the world—Kerry Emanuel of MIT and Ning Lin of Princeton—who published the most detailed hurricane model in history in August 2015. Emanuel and Lin's hurricane model was embedded within six different worldwide climate models routinely run by supercomputers.

"The term 'black swan' is a metaphor for a high-consequence event that comes as a surprise. Some high-consequence events that are unobserved and unanticipated may nevertheless be predictable," they wrote in *Nature Climate Change*. "Such events may be referred to as 'grey swans' (or, sometimes, 'perfect storms'). Unlike truly unpredicted and unavoidable black swans, which can be dealt with only by fast reaction and recovery, grey swans—although also novel and outside experience—can be better foreseen and systematically prepared for."[6]

Lin and Emanuel said their research showed that not only were grey swan hurricanes now likely to occur, one such devastating hurricane would almost certainly hit the Persian Gulf region—a place where tropical cyclones have never even been seen in history. They identified a "potentially large risk in the Persian Gulf, where tropical cyclones have never been recorded, and larger-than-expected threats in Cairns, Australia, and Tampa, Florida."

Emanuel and Lin showed that the risk of such extreme grey swan hurricanes in Tampa, Cairns, and the Persian Gulf increased by up to a factor of fourteen over time as Earth's climate changed.

"These are all locations where either no one's anticipated a hurricane at all, such as in the Persian Gulf, or they're simply not aware of the magnitude of disaster that could occur," Emanuel told reporters at the time.[7] He said that every hurricane that has ever occurred in recorded history could have been predicted (in retrospect) given the previous pattern of storm activity in that particular part of the world.

"In the realm of storms, I can't really think of an example in the last five or six decades that anybody could call a black swan," Emanuel said. "For example, Hurricane Katrina was anticipated on the timescale of many years.

Everybody knew New Orleans was going to get hammered. Katrina was not meteorologically unusual at all."

Still, the notion of predicting the growing likelihood of grey swans—which would look an awful lot like Category 6 hurricanes that don't exist now on the scale used to describe the strength of hurricanes—is something that no one talks about, but they can be imagined, perhaps even anticipated.

Masters drilled down into Tampa to make his point—and to illustrate just how devastating a grey swan hurricane would be in a city that has only seen two major hurricanes since 1848. In 2017, Hurricane Irma nearly became the third to hit Tampa. Only a last-minute course correction in the path of the hurricane kept it from hitting the city directly.

"Tampa Bay doesn't get hit very often by hurricanes," Masters wrote in his post.[8] "This is because the city faces the ocean to the west, and the prevailing east-to-west trade winds at that latitude make it uncommon for a storm to make a direct hit on the west coast of Florida from the ocean. This is fortunate, since the large expanse of shallow continental shelf waters offshore from Tampa Bay (less than 300 feet deep out to 90 miles offshore) is conducive for allowing large storm surges to build."

But should a grey swan hit Tampa, in a worst-case scenario, Masters said that the entire city would essentially be engulfed in a wave of water. The last big hurricane to hit Tampa (in 1921) produced a devastating storm surge roughly nine feet high. The Emanuel-Lin model showed that a storm surge of anywhere from twenty to thirty-six feet could hit Tampa.

The two hurricanes that have hit Tampa—in 1848 and 1921—were Category 1 and Category 3 storms. The 1848 hurricane generated a storm surge of fifteen feet in one instance, taking out much of what was in downtown Tampa at the time. A Category 5 storm—or, worse, something even stronger that looks like a Category 6 superstorm—would inflict untold damage.

Masters generated a map to illustrate what it might look like. "Downtown Tampa Bay would be inundated by more than 20 feet of water, and St. Petersburg would become an island, as occurred during the 1848 hurricane," he wrote. But that storm surge (twenty feet) might even be conservative.

A Category 6 storm could produce a storm surge of up to thirty-six feet, the models show. Such a monster storm is unlikely to hit Tampa, but it is also now not impossible given the changing conditions on Earth that are making mega-storms more powerful in intensity and strength.

"We might need to invent a 'Category Six' designation . . . that showed an unimaginably intense hurricane with . . . top sustained winds of 233 mph, traveling parallel to the coast along just the right track to generate a titanic 36-foot storm surge," he wrote. "Even accounting for the 15% reduction in winds that would occur due to friction over land, the winds from such a 'Category Six' hurricane would be like those of the EF5 tornado that leveled Joplin Missouri—except that EF4 to EF5 damage would be along a swath 22 miles wide, instead of a few hundred yards wide!"

In the event of such a storm, more powerful than anything Tampa has experienced, city officials may have no idea what they truly face. At least one city planning document (from 2010) anticipated that a Category 5 hurricane could cause two thousand deaths and $250 billion in damage. But it could be far worse.

"A storm surge of 5 meters is about 17 feet, which would put most of Tampa underwater, even before the sea level rises there," Emanuel told reporters. "Tampa needs to have a good evacuation plan, and I don't know if they're really that aware of the risks they actually face."[9]

A city like Dubai is even more unprepared, Emanuel said. Dubai, and the rest of the Persian Gulf, has never seen a hurricane in recorded history. Any hurricane, of any magnitude, would be an unprecedented event. But his models say that one is likely to occur there at some point.

"Dubai is a city that's undergone a really rapid expansion in recent years, and people who have been building it up have been completely unaware that that city might someday have a severe hurricane," Emanuel said. "Now they may want to think about elevating buildings or houses, or building a seawall to somehow protect them, just in case."

And Emanuel hinted that it might be time for meteorologists and others to alter their hurricane scale—in effect, to anticipate what a Category 6 hurricane would be. "You're going to see an increased frequency of the most extreme events," he said. "Whereas the upper limit of hurricane wind speeds today might be 200 mph, 100 years from now it might be 220 mph. That means you're going to start seeing hurricanes that you've never seen before."

Following Masters's provocative post, many of his meteorologist colleagues weighed in. The Weather Channel predicted that a Category 6 hurricane, and a change in the scale to accommodate it, may be on its way.

"Jeff Masters got the entire weather community thinking: Could there be a Category Six hurricane?" Brian Donegan wrote on the network's site. "Last year, Hurricane Patricia reached maximum sustained winds of 215 mph in the eastern Pacific Ocean. It was the most intense tropical cyclone ever recorded in the Western Hemisphere."[10]

Patricia was the fastest-intensifying hurricane in history, with the highest reliably measured winds ever observed on Earth over the ocean. A hurricane with winds that high—which will occur more often, with greater frequency now—has meteorologists wondering. Hurricane Patricia "featured winds well above the 155-mph criteria of a Category 5 hurricane. But should [it] be considered a Category Six?" Donegan asked. "The only way that is possible is if the National Hurricane Center decides in the future to adjust its Saffir-Simpson Scale."

A fellow meteorologist, Paul Huttner, said Patricia makes it all but certain that we'll see Category 6 hurricanes. "Many meteorological observers [were] stunned at how rapidly Patricia blew up from tropical storm to one of the strongest Category 5 hurricanes on earth in just 24 hours," Huttner wrote for Minnesota Public Radio.[11]

Huttner also asked, publicly, the question that other meteorologists have begun to ask privately in the wake of Hurricane Patricia, and a second hurricane (Joaquin), that spun up very, very quickly and with nearly as much intensity.

"Forecast models simply could not grasp the rapid intensification of Patricia," he wrote. "The recent rapid intensification of storms like Joaquin and Patricia over super-warm ocean water begs important questions. Are we entering an era of new atmospheric physics where traditional weather forecast models can't keep up with actual dynamic changes? Is the quickening pace of the hydrologic cycle too fast for today's forecast models to handle?"

It's a tremendously important question. Seven Category 5 typhoons hit during 2016, above the normal average of four. The Philippines saw two big tropical cyclones hit in just one week, which is also quite rare. Had any of these spun up overnight, or into mega-storms with ferocious, sustained winds above two hundred miles an hour, it's hard to even calculate the potential loss of life and property.[12]

Whether we call them Category 6 hurricanes—or simply Category 5 hurricanes with really fast, violent winds that are up to sixty miles an hour above the upper end of the current scale that can appear literally overnight over warm oceans—we need to be ready for these superstorms capable of taking out cities like Dubai or Tampa. They are here, right now. The devastation we saw in Houston, Puerto Rico, and the Caribbean in the fall of 2017 is a clear warning. We ignore the implications at our peril.

11

The Displaced

Aslow-moving storm began building over the Atlantic Ocean in early August 2016. It didn't appear to be about to turn into a fierce hurricane with gale-force winds, so no one outside the relatively small group of meteorologists who pay close attention to such things noticed the track of the storm.

The storm system was being fed by two things: near-record warm seas in both the North Atlantic and the Gulf of Mexico and the extra precipitation that now circulates in the atmosphere. Both of these things—really warm ocean water and extra precipitation capable of gathering regionally in certain parts of the world at any given time, depending on other factors—are new elements in a world that's changing in real time, with real-world consequences. It was precisely these same dual conditions that later fueled Hurricane Harvey, "adding to the rain that kept Houston underwater for days," and Hurricane Maria, leading "to floods and landslides in Puerto Rico," as well as the Atlantic's "unusually warm waters" that added to Irma's intensity in fall 2017 as it destroyed islands in the Caribbean.[1]

By the time the storm made landfall in the southeastern United States in 2016, a few weather forecasters started to pay close attention as well. But it was still a local story, and not especially out of the ordinary. People in Louisiana, who are accustomed to such storms and deluges, were prepared to ride out the heavy rain. Newscasters warned people to watch for flash floods, which is now standard practice during periods of heavier rainfall.

But this storm system was bigger than others, with more rain in store for every coastal city in its path. It would last for days on end, across nearly the entire southern coastline of the United States. From August 9 to August 14, the storm inundated Louisiana, Mississippi, Alabama, Florida, and Texas.

In fact, there was so much rain over the six days that people began to realize it was something out of the ordinary. Five cities in Louisiana reported rainfall totals of more than two feet. Some parts of the state saw more than twenty inches of rain in just one forty-eight-hour period. In one particularly strange turn of events, a well-known local river actually reversed course for a period of time.[2]

It was the type of rainfall event that meteorologists and atmospheric scientists say should only occur once every thousand years, based on what's happened historically. Thirteen people died in the flash floods that erupted in city after city across the southern coast of the United States. The American Red Cross said at the time that it was the "worst disaster since Superstorm Sandy [in New York City]."[3] Just a year later, they would need to revise that estimate when Hurricane Harvey leveled Houston.

But here's what's even more significant. The torrential rains that covered the American South in 2016 may have been a one-in-a-thousand-year extreme weather event, but so was Hurricane Harvey, which occurred only a year later. That same fall, Irma hit the Caribbean with record-breaking wind speeds of "180 mph for 37 hours, longer than any storm on Earth."[4] Shortly afterward, Puerto Rico was hammered with Hurricane Maria's 155 mph winds, destroying their power grid and creating a "massive humanitarian crisis on the island that could endure for months," said Elaine Duke, acting secretary of the Department of Homeland Security, in September 2017.[5]

With multiple "one-in-a-thousand-year events" occurring so close together, it is highly likely there will be more such extreme events—long, long before the year 3000. There may be another one in the next ten years . . . or next year.

Scientists are now able to closely study extreme events, such as the one that battered America's southern coastline for nearly an entire week during August 2016, and determine whether the changed system on Earth was, in fact, responsible for it. It's called "attribution science."

Five global research centers, working together, said that Earth's changed system had doubled the chances for the type of a one-in-a-thousand-year flood to hit in Louisiana and other states.[6] They discovered two important factors at play during the event. On August 11, a measure of atmospheric moisture (something that scientists call *precipitable water*) had reached a historic level of 2.78 inches.

That's a level higher than during some past hurricanes in the region. Extreme precipitation levels in the Southeast have increased 27 percent from 1958 to 2012. Over the past century, the United States has seen a 20 percent increase in the amount of precipitation falling in the heaviest downpours, which has dramatically increased the risk of flooding. Since the 1980s, a larger percentage of precipitation has come in the form of intense single-day events, and nine of the top ten years for extreme one-day precipitation events have occurred since 1990.

In short, the world has changed—even in the United States. We can expect more and more extreme precipitation events like what we saw in the Louisiana and Houston floods now.

"Before the floodwaters had receded, the media began questioning whether this extreme event was caused by anthropogenic climate change," the researchers wrote. "To provide the necessary analysis to understand [its] potential role . . . a rapid attribution analysis was launched in real time using the best readily available observational data and high-resolution global climate model simulations."

The August 2016 deluge, they concluded, can be attributed to Earth's changed system. In fact, the changing climate is now responsible for nearly 20 percent of extreme rainfall events. The more extreme the event (like the Louisiana floods), the more likely Earth's changing climate is responsible.

One of the clearest changes in weather globally and across the United States is the increasing frequency of heavy rain. A warmer atmosphere holds more water vapor. And, like a bigger bucket, a warmer atmosphere dumps more water when it rains. The storm in the southeastern United States was supercharged by running over a warmer ocean and through an atmosphere made wetter by global warming.

The Louisiana floods were the first high-profile event in America to create a deluge of another kind that will only become more common everywhere—environmental refugees. We can expect more of these, even in America. People in Alaskan cities are being forced to relocate. Extreme wildfires in California are driving people from their homes. A fire near Los Angeles almost at the same time as the Louisiana floods required more than eighty thousand to leave their homes as a precaution.

But the one-in-a-thousand-year flood in Louisiana was the first big displacement in America that truly caught people's attention in ways that hadn't before. More than forty thousand homes were damaged, forcing people to relocate, while thirty thousand people were rescued from flash floods and temporarily relocated as well. Houston, a year later, reminded us yet again what we may face now on a regular basis.

"I don't know [that] we have a good handle on the number of people who are missing," Louisiana's governor, John Bel Edwards, said at the time.[7] Twenty parishes were declared federal disaster areas. More than eighty thousand people registered for the FEMA Individual Disaster Assistance Program.

Louisiana may have been the first such extreme flooding event in the United States to create refugees, but it won't be the last. If one-in-a-thousand-year flooding events become common every decade or so, or even every few years, we will see lots and lots of these. All will create refugees who are fleeing not because of armed conflict or civil unrest but because extreme "natural" events have displaced them.

Louisiana also isn't alone. Extreme events are now creating refugees all over the world. Sometimes it's an extreme precipitation event in the Philippines that causes severe coastline damage. At other times, it's a drought that lasts for up to five years. In still other places, it's a typhoon that looks an awful lot like a Category 6 storm that flattens buildings in a very wide path.

And in some instances, it is severe water scarcity that drains nearly every freshwater aquifer in the country (Yemen) or food insecurity that sends millions to the brink of starvation (Somalia).

No one is prepared for the rising tide of environmental refugees that currently hide inside much larger numbers of migrants and refugees fleeing armed conflict in war-torn regions of the world. The United Nations has been singularly unsuccessful in recognizing the term under the humanitarian mission that has been at the heart of the global community since the Second World War.

People generally know what an "environmental refugee" is—in concept or theory, at least. So do various arms of the United Nations—again, in theory. For instance, the UN's Organisation for Economic Co-operation and Development (OECD) has an excellent definition for it. "An environmental refugee is a person displaced owing to environmental causes, notably land loss and degradation, and natural disaster," OECD says in its official glossary of statistical terms, which is supposed to guide how the UN body as a whole treats people who are defined by such terms.[8]

The International Organization for Migration has a straightforward definition as well: "Environmental migrants are persons or groups of persons who, for compelling reasons of sudden or progressive changes in the environment that adversely affect their lives or living conditions, are obliged to leave their habitual homes, or choose to do so, either temporarily or permanently, and who move either within their country or abroad."[9]

But these words, and definitions, don't actually matter to courts or the UN's humanitarian agency when it comes to the way in which they treat such refugees. Only victims of war, persecution, or armed conflict are awarded status and humanitarian relief.

The reason is that the UN's humanitarian treaty and charter was formed and ratified decades ago, long before the environmental landscape had begun to change in ways that created wholesale displacement of people as migrants and refugees. Millions of people are fleeing parts of the world that can no longer sustain them, but there is no legal mechanism for the UN and governments to respond as they do to other things like war or disasters.

Simply put, the United Nations' refugee agency, the UN High Commissioner for Refugees, doesn't recognize them. "They are not covered by the 1951 [refugee] convention," UNHCR's Marine Franck told Al Jazeera.[10]

The simple but sad truth is that environmental refugees don't fit into any of the legally recognized definitions of a "refugee" that guide humanitarian efforts and legal status. Many such refugees migrate from one place to another within their own countries. They don't qualify. The UN's 1951 charter only recognizes and protects refugees who leave their home because they're afraid of being persecuted or because of "generalized violence or events seriously disturbing public order."[11]

Despite word changes (like OECD's definition) and other efforts to expand the definition of who a refugee is, the plain fact is that people who are forced out of their homes, off their lands, or away from their homelands because of changes in the environment are still not offered the same legal protection as refugees who meet the 1951 charter's legal guidelines. It's an absurd situation—especially, as we've seen, when places like Yemen, Syria, Somalia, and others are seeing massive migrant movements well before civil unrest, persecution, or military conflict begins.

In 2011, experts warned at the annual meeting of the world's largest scientific organization (the American Association for the Advancement of Science, or AAAS) that there could easily be tens of millions of environmental refugees by the end of the decade. "We will have 50 million environmental refugees [by 2020]," said UCLA professor Cristina Tirado. "When people are not living in sustainable conditions, they migrate."[12]

Migrations from northern Africa to Southern Europe have been a terrible problem for years. A perfect example is Tunisia, where food shortages and widespread unemployment and poverty toppled the government of its longtime ruler, Zine El Abidine Ben Ali. That sort of political unrest will happen elsewhere.

"What we saw in Tunisia—a change in government and suddenly there are a whole lot of people going to Italy—this is going to be the pattern," Michigan State University professor Ewen Todd said at that same AAAS briefing.

Already, Africans are going in small droves up to Spain, Germany and wherever from different countries in the Mediterranean region, but we're going to see many, many more trying to go north when food stress comes in.

And it was food shortages that put the people of Tunisia and Egypt over the top. In many Middle Eastern and North African countries, you have a cocktail of politics, religion, and other things, but often it's just poor people saying "I've got to survive, I've got to eat, I've got to feed my family" that ignites things.

Experts have been debating the theory of environmental refugees for years. The Royal Society, for instance, began to debate it extensively a decade or so ago. It has regularly issued serious reports and papers on it over the years.

"Massive population displacements are now regularly presented as one of the most dramatic possible consequences of climate change," one typical Royal Society paper said in 2011.[13]

The concept, in theory, makes perfect sense to academic experts who study such trends. "These are people who can no longer gain a secure livelihood in their homelands because of drought, soil erosion, desertification, deforestation and other environmental problems, together with the associated problems of population pressures and profound poverty," says Oxford University's Norman Myers.[14]

Myers, who has been a vocal proponent of the theory of environmental refugees for years, has estimated that there could easily be 250 million such refugees in the near term in countries like China, India, Bangladesh, and most of the cities of the Middle East in and around the Horn of Africa.[15]

Theory is one thing. The reality is quite another, and humanitarian and world leaders haven't kept pace with it. The estimate of fifty million environmental refugees by 2020—first enumerated at the AAAS annual conference in 2011—could easily become five times this in the real world.

As one example, forty-two million people were displaced in Asia and the Pacific by storms, floods, and heat waves in 2010 and 2011 alone, according to one of the definitive experts on refugees (the Internal Displacement Monitoring Centre). While many eventually returned to their homes, millions did not and settled somewhere else inside or outside their country's borders.[16]

"When a cyclone destroys homes and people look for shelter elsewhere, it is clear that they were displaced by disaster," the Wilson Center said in a December 2015 special report. "But when long-term drought and repeated crop failures in a region like the Sahel in Africa deepens poverty among communities already marginalized and harassed by the state because of ethnic or political cleavages, climate, poverty, and persecution all play a role in the decision to leave. Attempts so far, however, to expand the legal definition of refugees to include those fleeing natural disasters or to revise the refugee convention, have gotten nowhere."[17]

Advocacy groups are starting to mount fierce campaigns to force the UN and others to finally recognize what is an obvious, new reality. The Natural Resources Defense Council, one of the oldest and best-known environmental organizations in the United States that regularly sends dozens of its lawyers into court, has taken just such a stance.

The UN's 1951 outdated charter, with its definition of refugees fleeing from war or persecution, is "incomplete and [has] the effect of locking some of the world's most indigent and at-risk people in their circumstances," NRDC said in a special report.[18]

"Those who lose their homes in earthquakes . . . and whose countries cannot help them rebuild are not refugees. People starving to death in abject poverty are usually not eligible to be refugees either, because so many of their countrymen tend to suffer the same way. Indeed, one of the tragic ironies of refugee law is that, as the number of people in trouble grows, the less likely any of them are to qualify for protection. The law is concerned only with persecuted minorities," it said.

Here's how absurd the situation is right now. Kiribati, an island in the Pacific Ocean, is physically disappearing into the sea. Saltwater is intruding into its soils and its drinking water. Kiribati's homes will be uninhabitable before too long.

So a fisherman in Kiribati—Ioane Teitiota, who makes his living from the sea—tried to move to New Zealand as a refugee to make a new home for himself. New Zealand's highest court rejected his petition to relocate as a refugee, because there was no legal basis for it, and sent him back to Kiribati in September 2015.[19]

The New Zealand judicial ruling was harsh. "Traditionally a refugee is fleeing his own government or a non-state actor from whom the government is unwilling or unable to protect him," the New Zealand judges wrote. "The claimant is seeking refuge within the very countries that are allegedly 'persecuting' him."[20]

Besides the fact that this is a tortured, ridiculous framework for a legal ruling, the judges did, in fact, address what they and many others in countries like New Zealand are now afraid of—namely, that granting the Kiribati fisherman's request for asylum would likely open the floodgates for potentially tens of millions of migrants to seek asylum in more hospitable, sustainable parts of the world.

"At a stroke," the judges on New Zealand's High Court of Appeals wrote, "millions of people who are facing medium-term economic deprivation, or the immediate consequences of natural disasters or warfare . . . would be entitled to protection under the Refugee Convention."

New Zealand isn't alone. The United States, under President Trump, is closing its doors to refugees. So are lots of other countries. The last thing any of them want is another basis on which people can seek asylum—even when it's patently obvious that they are leaving one place for another because their environmental landscape has changed and no longer supports them, their neighbors, or their communities.

But as we've seen in country after country—from Yemen's water riots and Syria's farm migration to urban centers in the face of horrific drought to millions of Somalians facing starvation over food security and the Pakistani people living in fear of a decision by India to close off their access to fresh water—the things that make up the fabric of what a refugee or migrant look like are intertwined.

Diplomats knew civil war was coming in Yemen long before it actually erupted because the country had largely run out of fresh water. People fleeing Yemen because they have no water to drink are refugees—just like those very same people who later flee Yemen after civil war destroys their country and their government.

It will become much more difficult for leaders of countries that genuinely care about humanitarian relief efforts to deny this reality for much longer. However, until the United Nations accepts "environmental refugees" as a category, such refugees will only be people who live in places where the land no longer sustains them and who desperately want nothing more than to escape their circumstances. Only when their governments choose to persecute them or put them in harm's way can they officially become "refugees" in status, which means they can then receive humanitarian assistance.

In the meanwhile, they are stranded in a political no-man's-land that cannot help but breed either depression or rebellion.

12

Vanishing Icons

In late spring of 2015, a BBC camera crew accompanied a research team to the remote steppes of central Kazakhstan to film an extraordinary event in the animal kingdom.[1] Tens of thousands of female saiga were gathered to give birth on the open plains over the course of just ten days—as they'd done every spring since the Ice Age. It's one of the most spectacular migrations and mass birth events in the world.

The saiga, an antelope that has survived for millennia and once ran with the wooly mammoths, give birth in mass numbers in open spaces in order to offer their calves the best chance for survival. Wolves, the saiga's main predator, can only capture and kill a relatively small percentage of the calves in such a mass-birthing process. Like horses, which were also once flight animals in the wild until they were largely domesticated, saiga antelope calves are born quite large and well developed and fully capable of outrunning predators within a few days of birth.

Once, hundreds of thousands of saiga roamed the plains of Russia and other countries within the former Soviet Union. Over time, however, the saiga had been hunted and poached almost to the point of extinction. As the Soviet Union collapsed, hunters killed saiga antelope in record numbers for meat, while poachers killed males of the iconic species for their horns (used in traditional Chinese medicine).

By the first decade of the twenty-first century, the saiga antelope's numbers had diminished worldwide to a mere fifty thousand. The animal was listed on critically endangered lists. But an intensive effort by wildlife conservation and saiga research scientists—including several who accompanied that BBC crew in the spring of 2015—had successfully brought the saiga population back to healthy, annual numbers. By 2015, the saiga population in the world had stabilized to about three hundred thousand.

So the research team that came to the plains of Kazakhstan fully expected to witness a mass-birthing event that rivaled those of years past, when saiga were plentiful and roamed widely across the grasslands. What the team witnessed instead was the mass die-off of an estimated two hundred thousand saiga almost overnight. The corpses were littered over hundreds of miles. Every single animal died over a period of several days.

The mass die-off was captured by the BBC team and transfixed global media coverage for several days. The images of the saiga females' and calves' corpses strewn across the open plain in Kazakhstan were hard to ignore. They were everywhere in the media during the die-off.

What no one could answer was why it had happened. Mass die-offs occur every so often among an animal species. Saiga had gone through die-offs on a smaller scale in the past, but earlier die-offs had been local and could be traced either to a toxin introduced into the local environment where they had died or to an infectious disease that had taken weeks or months to cause the mass deaths.

In this case, in Kazakhstan, two hundred thousand saiga died over a relatively broad area . . . and essentially overnight. Infectious disease made no sense (because it happened so quickly). A toxin made no sense (because the die-off wasn't local or contained).

So what caused it? Why did this mass death happen? That was the question that researchers, conservation societies, and journalists asked for months following the die-off, with no good answers. There were crazy speculations, ranging from the Blood Moon and aliens to signs of an imminent apocalypse.[2]

Finally, a year after the event—as a vastly diminished herd of saiga were gathering on the plains again for another mass-birthing event—at least a partial answer emerged from accelerated research and studies. The Saiga Conservation Alliance had sent a research team in and had analyzed tissue samples for months.

"The mysterious death of 200,000 critically endangered antelopes in Kazakhstan last year was caused by a bacterial infection, according to a new report by the Saiga Conservation Alliance," CNN reported on April 15, 2016, nearly a year after the spectacular mass die-off.[3]

"Several labs used tissue samples collected from the carcasses during the die-off, and confirmed that the deaths were linked to bacterium *Pasteurella multocida*. This pathogen caused hemorrhagic septicemia in the saiga population," CNN reported. "The findings are surprising because this bacterial infection, although known to affect animals such as buffalo, cattle and bison, had never been documented to affect an animal group with a 100% mortality rate."

Every other news report said essentially the same thing. Something had triggered the bacteria, which is usually dormant and not lethal. The bacteria had somehow become deadly to every saiga across the entire steppes. And this virulent bacterial infection, which strikes small percentages of grasslands creatures like buffalo and cattle, had managed to afflict a "100% mortality rate."

None of the follow-up media stories addressed still-unanswered questions. Why a 100 percent mortality rate? Why did these usually harmless bacteria become virulent? Was there a trigger of some sort—either an environmental change or an internal change in the saiga's immune system that made it susceptible to the bacteria? Or was it a combination of both?

Several of the initial researchers involved with the conservation efforts surrounding the saiga and the first BBC filming of the mass die-off in the spring of 2015 that stunned the world now have a more definitive answer. They talked about the underlying cause in a first-person account in *The Conversation* in December 2016 that explains in some depth what went wrong in the plains of Kazakhstan.[4]

"In exploring these questions, our research is a Russian doll; as we take off a layer of explanation we find more questions within," wrote E. J. Milner-Gulland, Eric Morgan, and Richard Kock. Milner-Gulland is the chair of trustees of the Saiga Conservation Alliance that reported on the proximate cause of the die-off a year after it had occurred. Morgan and Kock also received research funding to study the saiga event.

Once they'd peeled back the layers, the underlying causes started to become clear. They aren't especially pleasant answers, but they should also give us pause, because they give us a vivid snapshot of similar events that will become more prevalent over time—not less—as Earth's climate system changes. And future events might not always kill off only a particular animal species. Bacteria that can flip from dormant and harmless into virulence can also affect humans.

"We [went] back to old field notes from the Institute of Zoology in Kazakhstan for 1988 when a similar mass mortality occurred; reviewed research on mass deaths in other species; looked for differences in the vegetation composition between the 2015 die-off and in other years; and built statistical models to explore changes in temperature and rainfall over a range of different temporal and spatial scales," the three researchers said.

"We also tested tissue and environmental samples for a wide range of toxins, as well as other disease-causing agents, in case some underlying infection was involved. So far, the evidence points towards a combination of short-term but landscape-scale weather variation and physiological stress from calving causing a cascading effect of virulence. There's no evidence for environmental toxins, other underlying infections or (as yet!) alien influence."

The complex answer, it seems, is that the combination of temperature changes combined with extreme and erratic precipitation—both signatures of the new Earth system that will affect regions of the world more frequently from here on out—had the net effect of creating a "short-term but landscape-scale weather variation." This change in the total environment, combined with physiological stress from calving, caused the ever-present bacteria in the saiga to become virulent and kill them all.

The saiga are prone to disease and birth-related mortality. What triggered the "100% mortality rate" in this case was a new climate system that created a temporary weather variation. This sort of thing is going to happen again, and more frequently, the researchers warned.

"There has been huge public interest in this event, both within Kazakhstan and globally," the researchers said. "People want quick answers and they want us to find solutions so that this will never happen again. It seems, however, that we won't be able to give the comfort that is wanted; in fact, it is likely that with climate change these types of events will become more rather than less prevalent."

The research that they were pointing to is an important study in *Science* from the summer of 2013 that firmly linked climate and public health threats like infectious diseases.[5]

"The life cycles and transmission of many infectious agents—including those causing disease in humans, agricultural systems, and free-living animals and plants—are inextricably tied to climate," ecology and evolutionary biologists from five American universities reported. "Over the past decade, climate warming has already caused profound and often complex changes in the prevalence or severity of some infectious diseases."

The world has already begun to change, they said, and the public health risks are starting to increase. "Climate change has already increased the occurrence of diseases in some natural and agricultural systems."

While experts continue to debate how swiftly it will cause serious harm to humans, the impacts on animals are more clearly understood now. Animals are, in effect, harbingers of our own future.

"For example, although the effects of climate warming on the dynamics of human malaria are debated, climate warming is consistently shown to increase the intensity and/or latitudinal and altitudinal range of avian malaria in wild birds," they reported.

The saiga antelope die-off was a highly publicized example of this new threat, layered on top of other stresses on the species. It's a complicated story, with answers that have taken nearly two years to sort through. The good news for the saiga is that they have begun to repopulate. Their numbers doubled, to about one hundred thousand, in 2016.[6]

But the overarching threat that created a "100% mortality rate" event for the saiga antelope in 2015 is with us now and will only increase. Events like the saiga die-off will happen again and more frequently. That's what the study in *Science* shows. It sends a clear signal to every species on Earth.

What the saiga story also illustrates is just how difficult it is to sort through myriad threats to species—including humans—even in simple ecosystems such as the open grasslands of Kazakhstan. Explanations for threats to iconic species like the saiga antelope are never simple, including poaching, land development, and, at present, a changing environmental landscape they may not be equipped to handle.

The extinction-level threats facing iconic species like the African elephant, the giant panda in China, African giraffes, or mountain gorillas in Africa are good illustrations. These species are in serious trouble. All face a multitude of dangers, some more widely known than others. Cutting across them sharply now is an environmental landscape that is shifting under their feet into something they simply may not be able to adapt to.

Mountain gorillas, for instance, are critically endangered. There are fewer than 900 of them currently in the wild, living in two national parks in central Africa. The species was discovered by us just a century ago. Since then, through a combination of uncontrolled hunting, war, disease, the destruction of its forest habitat, and illegal capture for trade, we decimated the species. The number of mountain gorillas fell to a low of just 620 in 1980 before conservation efforts brought the numbers back slightly, but the species' newest threat—a changing Earth system that systematically conspires against them in ways that other threats do not—may be the most difficult.[7]

Human beings, driven by water scarcity and rising temperatures, are encroaching on the gorillas' habitat. In northern Rwanda, for instance, people routinely head into the national park for water. Warming temperatures are pushing the mountain gorillas to higher elevations. As people edge farther into their habitat, mountain gorillas are driven even higher into the mountains, where they can suffer from exposure to harsher temperatures and find less to eat. While it's unlikely that mountain gorillas will simply vanish for good—thanks to concerted efforts by wildlife conservation societies—the environmental threat is real and ever present.[8]

Giant pandas, like mountain gorillas, are still with us now largely because of conservation efforts and zoos. The giant panda population, which is native to central China, has largely been decimated by habitat loss from logging, agriculture, and infrastructure expansion. What's left of the pandas' habitat is now protected through a network of nature reserves.

But in an unprotected world, pandas would likely vanish as a species over time. They have a finicky digestive system. They have to eat anywhere from twenty to forty pounds of bamboo a day to stay alive. A study several years ago, however, found that up to 100 percent of the bamboo in pandas' native habitat was under threat from the changing climate—meaning that pandas literally would have nothing left to eat if they weren't being protected in nature reserves.[9]

African elephants are the largest species walking on Earth. They wander through nearly every country in Africa. There were once ten million elephants in Africa before human colonization. Now there are only 350,000 of them left, with an ongoing species rate decline of 8 percent a year. Some experts now genuinely fear that African elephants, one of the continent's most iconic species, could disappear in a generation.[10]

African elephants are now vulnerable to extinction for many of the same reasons that other big, iconic species are in serious trouble—uncontrolled hunting, habitat loss from development, poaching, and disease.

The changing landscape in Africa is a grave threat to them. Elephants' DNA varies only modestly and adapts to environmental changes slowly over long periods of time. For this reason, most experts believe it may not be able to adapt in time to the changes that are beginning to transform Africa.

But the biggest problem facing the African elephant is precisely the same problem that is confounding human populations across the continent—water scarcity.

"The biggest concern for elephants is their need for large amounts of fresh water, and the influence this has on their daily activities, reproduction and migration," WWF said in a special report on the extinction threat facing African elephants. "Other threats—like poaching, habitat loss and human-elephant conflict—remain, and have the potential to increase due to the effects of climate stressors on humans and resulting changes in livelihoods."[11]

Their cousin, the Asian elephant, faces similar stresses. They require extensive amounts of water to live—up to sixty gallons a day. For this reason, Asian elephants tend to remain close to freshwater sources. As these are beginning to dry up in the thirteen countries where roughly fifty thousand Asian elephants now roam, they will find it hard to adapt. And like their African cousins, they, too, adapt slowly to changing landscapes and will likely run out of time. The habitats of Asian elephants are just 15 percent of their historical range today.[12]

Perhaps the most surprising recent development was the announcement in December 2016 that giraffes in Africa were undergoing a "silent extinction" through a combination of poaching and the rapid expansion of farmland. Africa's changing landscape could finish them off, conservationists warned.[13]

Virtually unnoticed, giraffe numbers have plunged by up to 40 percent since the 1980s, the International Union for Conservation of Nature reported in its annual Red List of endangered species. The overall population of the world's tallest land animal had fallen below 100,000 (from roughly 163,000 three decades earlier).

It was the first time the group had ever listed the giraffe as vulnerable to extinction. The reason, it said, was because the giraffe's steep decline in numbers in sub-Saharan Africa had largely gone unnoticed.

"Whilst giraffes are commonly seen on safari, in the media and in zoos, people—including conservationists—are unaware that these majestic animals are undergoing a silent extinction," said the group's giraffe specialist, Julian Fennessy.

Other conservation experts swiftly weighed in on the sobering news. The same water scarcity and food security problems that were causing such troubles for humans in sub-Saharan Africa were likewise causing giraffes serious trouble. That, combined with war and civil strife, was trapping giraffes inside a tightening noose.

"People are competing for fewer and fewer resources and the animals are worse off . . . especially with civil strife," Craig Hilton-Taylor, who runs the group's Red List, told Reuters. He said that drought and Africa's changing landscape were making things even worse for giraffes.[14]

Other iconic species that schoolchildren read about in countless stories and tales are in serious trouble for all the same reasons—and serve as warning flares for us. The African lion population, for instance, has now fallen 90 percent.[15] Western black rhinos are already extinct. Experts believe that the entire population of wild rhinos are likely to go extinct in the next fifteen years.[16]

And a surprising study published in the *Proceedings of the National Academy of Sciences of the United States of America* in December 2016 showed that cheetahs—the fastest land animal on Earth—likewise simply haven't been able to escape the same multitude of threats facing other iconic species.

There are now only seven thousand cheetahs left in the world—down from a population of one hundred thousand a century ago. Habitat loss is mostly to blame for the cheetahs' demise, the researchers said, as they've lost more than 90 percent of their historical range. The changing landscape and climate system in Africa could finish the job.[17]

Whether the vanishing of iconic animal species, a dome of heat–killing waves, Category 6 superstorms, or extreme weather events that create a new and yet unrecognized category of refugees, all are realities of climate change and real-world consequences we cannot ignore.

PART 4

The Geopolitics

Critical changes are unfolding right now, across every continent. They have everything to do with threats to our global security.

Why should we care about water scarcity, food shortages, extreme weather events, and multiyear droughts in places far removed from us, such as Yemen, Syria, Pakistan, and China? On top of humanitarian concerns, environmental changes also threaten the geopolitical landscape due to conflicts over resources required to sustain human life.

Without water, no civil society is possible. People become desperate. A lack of fresh water can throw a country into civil war and destabilize an entire region, creating a breeding ground for worldwide terrorism. Yemen is proof of that. So is Syria, whose farmers—refused international aid during a lengthy and severe drought—left their farms, moved to major urban areas, and joined the rebellion against Assad.

"The Geopolitics" explores the critical changes in water economy for water-stressed countries such as Saudi Arabia, Yemen, Syria, Jordan, Somalia, Pakistan, India, and China. It also explains why climate change is pivotal to international security. With drought and starvation killing millions and prompting those remaining to migrate, rising sea levels forcing refugees to flee their native lands, and increased violence and warfare over shrinking freshwater resources, climate refugees will increasingly impact global stability and force immigration policy changes.

The real stories in "The Geopolitics" are clear warning signs that what has happened will happen again—and on an even larger scale—if we don't proactively address the root issue of water scarcity.

13

Seeing Around Corners

Early in 2016, at the start of the last year of Barack Obama's presidency, the second-in-command at the Pentagon, Robert Work, signed a directive that was sent to every American combat field commander in the world.[1]

No one outside the military and national intelligence community paid any attention to the directive at the time. The media, consumed by a presidential campaign unparalleled in the history of politics, ignored it.

The field directive—now in force as a prime directive for combatant commands around the world for at least the next ten years—required that everyone from the secretary of defense and the Joint Chiefs of Staff all the way down to field commanders place climate stresses like water scarcity, food shortages, and multiyear droughts in subtropical regions like Nigeria, South Sudan, Yemen, and Somalia front and center in their strategic planning.

"All DoD operations worldwide must be able to adapt current and future operations to address the impacts of climate change in order to maintain an effective and efficient U.S. military," the directive read.

Under it, military field commanders are now required to "incorporate climate change impacts into plans and operations and integrate DoD guidance and analysis in Combatant Command planning to address climate change-related risks and opportunities across the full range of military operations, including steady-state campaign planning and operations and contingency planning."

Military field commanders have to "assess the risks to U.S. security interests posed by climate change within their areas of responsibility and resulting risks to U.S. operations, and identify associated U.S. military capability gaps and operational risks." That includes assessing threats from natural resource constraints on a country-by-country basis.

There are several extraordinary things about this Pentagon directive, which directs field commanders in theaters of military conflict to finally start to look around corners at emerging resource threats to stability in countries like Yemen, Pakistan, and Syria. The twelve-page directive is basically their theater operations field manual.

For years, the national intelligence community and military leaders in the United States had struggled with the immediacy of the threat that Earth's changing system posed for American troops in conflicts in the Middle East, Africa, and Southeast Asia.

The CIA, Pentagon, and others had tried for years to engage the science community in ways that might help them identify immediate threats and imminent conflicts over resources like water or food. They desperately wanted to look around corners and understand when and how such conflicts would erupt in places like Yemen and Syria before they actually did.

"Right now there's a gap between, OK, we can have a weather forecast for what the weather's going to be in the next month, and then we have the climate forecast, which is 30 to 100 years out," an unnamed Pentagon official told a reporter several years ago as efforts to use science in the service of intelligence gathering were ramping up. "It really doesn't help the combatant commanders plan their operations."[2]

But their efforts to work with the science community over the years had stalled, failed, or were halted by Republican Party leaders in Congress who steadfastly opposed "soft" efforts like humanitarian aid in regions of conflict or science-based intelligence gathering that anticipated threats rather than sending soldiers, tanks, and planes in to combat them when they happened.

On the science side, the CIA set up a classified climate center in 2009 to assess both near- and long-term threats from coming wars over natural resources. Republicans immediately attacked it and threatened to block funding for it.

The only time the head of the secretive CIA center ever talked publicly was on background to two graduate students—Charles Mead and Annie Snider at the Medill National Security Reporting Project at Northwestern University—who wrote about its launch and why the CIA's leadership felt it was important at the time.[3]

The CIA senior analyst told Mead and Snider that he'd sat at his desk while torrential rains had flooded Pakistan—the worst natural disaster in Pakistan's history—and realized that it was a warning to the national intelligence community.

"It has the exact same symptoms you would see for future climate change events, and we're expecting to see more of them," the senior CIA analyst said. "We wanted to know: What are the conditions that lead to a situation like the Pakistan flooding? What are the important things for water flows, food security radicalization, disease and displaced people?"

Three years later, his CIA center was gone. Republican members of Congress mounted a sustained assault on it for years during appropriations battles.

"The CIA's resources should be focused on monitoring terrorists in caves, not polar bears on icebergs," Senator John Barrasso (R-WY) said when the CIA center was announced. The former CIA director Leon Panetta fought for years to keep the center open but closed it in 2012 under continued and withering pressure from leading Republican members of Congress.

The CIA also restarted a program in 2010 that allowed dozens of civilian scientists with security clearance to access classified satellite surveillance photos in order to detect crop failures in countries like North Korea that could lead to food shortages.

The program (Measurements of Earth Data for Environmental Analysis, known as MEDEA) was active in the 1990s, shut down during the administration of George W. Bush, and then restarted during the Obama administration.

This program was shuttered in 2015, also under pressure from Republicans in Congress who wanted the Pentagon to spend its $600 billion fighting wars rather than understanding and anticipating why and how they might happen.

Former president Obama, for his part, finally came to understand that wars and conflicts in places like Yemen, Syria, South Sudan, and Pakistan were now as much about issues like growing water scarcity; extended, multiyear droughts; and extreme or unpredictable weather events as much as political considerations—all things that scientists have studied and predicted for years.

The subtropics were clearly starting to see impacts—which would naturally and quickly lead to internal strife as they had in Syria and Yemen—and intelligence and military strategists needed to see this landscape clearly.

"ISIS is not an existential threat to the United States," Obama said toward the end of his administration. "Climate change is a potential existential threat to the entire world if we don't do something about it."[4]

While he couldn't keep Congress from shuttering both Panetta's CIA climate center in 2012 and the MEDEA program several years later over constant threats to defund them, Obama could require Pentagon leaders to take emerging resource conflicts into account when planning for military operations in the field. So he did.

The twelve-page Pentagon field directive issued in early 2016 represents a profound change for both military and national intelligence experts. After years of struggling to assess the real nature of the threats and security risks posed by a natural world that was changing quite rapidly, the Pentagon's combat commanders now had a directive that required them to take things like water scarcity in Saudi Arabia or food shortages in Egypt into account when planning for theater operations. Water riots in Yemen years before a civil war broke out could be anticipated, not managed after the fact.

"You can't be on the ground in Asia, Africa, or the Middle East and not see what's happening," Christine Parthemore, a former Pentagon official who now serves as the executive director of the Center for Climate & Security, told *Foreign Policy* magazine. "I think that is why we've seen so many defense, intelligence, and diplomatic leaders start growing concerned about the security implications of climate change far earlier than our political leaders, academic researchers, or the general public."

A few courageous military leaders over the years have said publicly what most of their colleagues would only acknowledge privately—that the natural world was morphing so rapidly that it was creating threats from wars over resources. Admiral Sam Locklear, for instance, said publicly that increasing threats from famines, extreme typhoons, and the loss of fishing habitats in poor fishing communities presented the biggest U.S. security risk in the Pacific Command that he directed.

"President Obama has it right. I think it is a significant existential threat," Locklear told *Foreign Policy*. While ISIS, Russia, and China are all issues that Pentagon leaders have to deal with, "you've got to stay focused on the big-picture issues that face humanity and their implications."[5]

In their in-depth reporting for the Medill National Security Reporting Project in 2011, Mead and Snider found that military and intelligence analysts at the time were simply not attuned to what seems so obvious to the rest of us now—that as predictable river flows, crop yields, and agricultural production go away, war and military conflict follow right behind within a matter of months or years.

The two graduate students interviewed dozens of government officials for their report and reviewed two decades of government reports on the nexus between natural resource constraints and the potential for military conflict. What they found was that virtually no one was prepared for a rapidly morphing natural world that would make resource wars inevitable.

"The U.S. government is ill-prepared to act on climate changes that are coming faster than anticipated and threaten to bring instability to places of U.S. national interest," Mead and Snider concluded.[6]

The problem was that climate projections were far too general and that scientists were focused on events that might happen well into the future—even as extreme events and conditions like prolonged drought and water scarcity in the subtropics on either side of the equatorial band were starting to occur regularly just as climate models had been hinting at for years.

"Climate projections lack crucial detail . . . and information about how people react to changes—for instance, by migrating—is sparse," Mead and Snider wrote. "Military officials say they don't yet have the intelligence they need in order to prepare for what might come."

A CIA veteran who had wrestled with this very problem for more than two decades was willing to talk to them, on the record, about the problem. Rolf Mowatt-Larssen, who also led the Department of Energy's intelligence unit from 2005 to 2008, told them that the national intelligence community was quite adept at stealing secrets and assessing military and diplomatic crises, but they were woefully behind in assessing how resource wars were just over the horizon in places like Yemen.

"I consider what the U.S. government is doing on climate change to be lip service," Mowatt-Larssen told them. "It's not serious."

It wasn't for lack of trying. The first attempt at the MEDEA program in the 1990s yielded useful strategic information on North Korean crop yields, for instance, but every step forward often led to two or three steps backward. The intelligence community invariably reverted to tried-and-true strategy assessments that relied almost exclusively on military and diplomatic movements rather than the drivers behind those actions in the first place.

General Michael Hayden, who led the CIA from 2006 to 2009, said that energy and water issues made President Bush's daily intelligence briefings, but the genesis of the resource conflicts that were creating economic, agricultural, and environmental instability in hot spots never made their way into them.

"I didn't have a market for it when I was director," Hayden said. "It was all terrorism all the time, and when it wasn't, it was all Iran."

While the CIA's centers and research arms in this area have been shuttered and there is no single center anywhere in the U.S. government charged with monitoring emerging threats from conflicts over natural resources like food and water, military and intelligence analysts have nevertheless begun to assimilate these emerging threats into their planning. The 2016 directive to combatant commanders is a perfect example of this. They know what's coming, and they don't want to fly blind.

Military experts from around the world gathered at the Chatham House at the end of 2016, for instance, and shared strategic assessments with each other in an off-the-record setting. People who attend Chatham House events such as this can discuss controversial or difficult subjects privately but cannot mention publicly what was said or shared outside the meeting.

"The US Department of Defense regards climate change as an 'accelerant of instability and conflict,'" Chatham House said in its overview of the meeting that was closed to media and outside observers. "A former head of the US Pacific Command described it as the most significant long-term security threat in his region. US federal agencies have recently been mandated to fully consider the impacts of climate change in the development of national security policy."[7]

The Chatham House overview said that "this step-change in the US approach reflects the Pentagon's conclusion that climate impacts are a 'threat multiplier' for security concerns—not just for the future—but which pose 'an immediate risk to national security.'"

A few of the participants at that Chatham House event were willing to characterize some of their conclusions in statements afterward. One of them said that Earth's changing system, combined with other factors, was driving tens of millions of people to despair in parts of the world and that it was only a matter of time before it led to humanitarian crises not seen before.

Just months later, in early 2017, the UN's top humanitarian official told the National Security Council that drought and starvation could kill as many as twenty million people in Somalia, South Sudan, Nigeria, and Yemen without intervention. It was the greatest humanitarian threat confronting the UN community since the Second World War, he said.

In the late 2016 Chatham House meeting, Brigadier General Stephen A. Cheney, the CEO of the American Security Project, said that forced migration from parts of the world like North Africa and the subtropics where water and food constraints are approaching catastrophic proportions is now the "new normal" in these parts of the world.

"Climate change could lead to a humanitarian crisis of epic proportions. We're already seeing migration of large numbers of people around the world because of food scarcity, water insecurity and extreme weather, and this is set to become the new normal," General Cheney said. "Climate change impacts are also acting as an accelerant of instability in parts of the world on Europe's doorstep, including the Middle East and Africa. There are direct links to climate change in the Arab Spring, the war in Syria and the Boko Haram terrorist insurgency in sub-Saharan Africa."[8]

Another speaker at the Chatham House event, Major General Munir Muniruzzaman, a former military adviser to the president of Bangladesh and later the chair of the Global Military Advisory Council on Climate Change, said that climate change in South Asia could lead to refugee problems on an "unimaginable scale."

"Bangladesh is . . . ground zero of climate change. With a one meter sea-level rise, Bangladesh will lose about 20 percent of its landmass. We're going to see refugee problems on an unimaginable scale, potentially above 30 million people. India has already unilaterally fenced in Bangladesh on all three sides, so if we see climate refugees in these numbers, it will destabilize not only Bangladesh, but also severely impact international stability in the region."

Muniruzzaman was blunt about the immediate future in China and countries around it. "South Asia is one of the most water stressed regions of the world. A combination of water scarcity in the climate induced conditions and regional politics has made the right brew for a potential conflict. South Asia therefore could be the scene of the first water war in the climate stressed conditions."

Rear Admiral Neil Morisetti, a former commander of the UK's maritime forces, said that the obvious instability risks are here, now. "Climate change is a strategic security threat that sits alongside others like terrorism and state-on-state conflict, but it also interacts with these threats. It is complex and challenging; this is not a concern for tomorrow, the impacts are playing out today."

The migration problem is now growing exponentially year after year— driven to a great extent by environmental factors. More than two hundred million people around the world were displaced by natural disasters between

2008 and 2015, and the risk has doubled since the 1970s, according to the Norwegian Refugee Council's "2016 Global Report on Internal Displacement."[9]

Most of the displacement takes place inside the boundaries of each country, but those driven across borders are not considered refugees, because the 1951 Refugee Convention recognizes only people fleeing war or persecution.

"The international system is in a state of denial," Bangladesh's Muniruzzaman said in January 2017 when international leaders met in his country's capital to attempt to fashion a new global migration policy. "If we want an orderly management of the coming crisis, we need to sit down now—we should have sat down yesterday—to talk about how the management will take place."[10]

Bangladesh is at the center of much of the discussions about the lack of a global migration policy or guidelines. It is likely to see more than twenty million people displaced in the next twenty years or so. Nearly all of them will migrate to the capital, which means that it will grow from fourteen million to forty million people in rapid fashion. Bangladesh's cities simply can't handle that sort of change that quickly.

"The settlement of these environmental refugees will pose a serious problem for . . . densely populated Bangladesh and migration [abroad] must be considered as a valid option for the country," the government said in an "action" plan[11] describing the problem. "Preparations in the meantime will be made to convert this population into trained and useful citizens for any country."[12]

Muniruzzaman told leaders at the conference that the government of Bangladesh was simply not prepared for what was about to come. "It could destabilize the country and it could also go to the point of state collapse," he said.

Bangladesh is a microcosm of what could likely happen globally as more and more refugees pack tightly into the same space, competing for the same resources needed to sustain life.

14

The Water Economy

Nestlé, 150 years old in 2017, is the largest food company in the world. It operates hundreds of factories and employs hundreds of thousands of people in nearly two hundred countries. It makes a wide diversity of products ranging from baby food and breakfast cereals to coffee and dairy products that depend on water—sometimes lots and lots of water. American consumers who buy Nestlé products tend to know them by brand names like Stouffer's and Gerber under the corporate umbrella. In addition, as part of the Global 500 companies, Nestlé "likes to call itself a leader in nutrition, health and wellness (not just a Big Food player)," *Fortune* reports.[1]

As Nestlé is one of the world's largest publicly traded companies, its senior executives must worry in great depth about threats to its global business empire. In some parts of the world, Nestlé senior officials pay close attention to political regime changes that might affect their business operations inside the country. In other parts of the world, they monitor environmental or ecological challenges.[2]

On a global basis, because they are a public company required to report material threats to their far-flung business empire, Nestlé officials track and forecast future threats to their businesses like weaknesses in global trade or the financial markets. Every big, successful multinational company does this. They report some of these emerging threats to shareholders, but much of it remains inside corporate headquarters as vital information needed to drive important business decisions.

In early 2009, several U.S. State Department officials from the U.S. embassy visited Nestlé's corporate headquarters in Switzerland for a briefing on matters of joint concern and interest to both Nestlé and the United States.

The embassy officials were stunned by the briefing they received from Nestlé's officials, which was reflected in the headline on the top of the secret diplomatic cable they sent back to the State Department in Washington, D.C.: FORGET THE GLOBAL FINANCIAL CRISIS, THE WORLD IS RUNNING OUT OF FRESH WATER.[3]

Ordinarily, such a meeting between U.S. embassy officials and senior officers at an enormous global company such as Nestlé might be reflected in a carefully worded public statement about the nature of the briefing. It's exceedingly rare that raw, unfiltered assessments from a briefing like this are ever publicly disclosed in real time. Virtually all such diplomatic cables remain secret for decades and are only assessed by historians long after all of the authors and participants named in the cables are retired, dead, or out of power.

But in 2010, WikiLeaks released hundreds of secret diplomatic cables it obtained by hacking into the State Department's cyber network. *The New York Times* and other news organizations published dozens of breaking stories about the WikiLeaks hack into then secretary of state Hillary Clinton's department.

The stories at the time ranged from explosive topics like a nuclear standoff with Pakistan; the potential for a regime collapse in North Korea; negotiations over the fate of the Guantanamo Bay prison; China's hack of Google's computer system; arms deliveries to the militant Hezbollah in Lebanon; and Afghanistan political corruption.[4]

It was the beginning of the long-running feud between WikiLeaks founder Julian Assange and Mrs. Clinton that later played a prominent and central role in the 2016 presidential elections. The WikiLeaks hack of the State Department cables, then and now, had a profound impact on U.S. diplomatic relationships across the globe.

Not surprisingly, no one paid any attention whatsoever to the relatively calm and seemingly benign Nestlé diplomatic cable buried inside the hundreds of other State Department cables released by WikiLeaks. It was, after all, just a report on an internal briefing—albeit an extraordinarily candid, unfiltered one—from senior officers at the world's largest food company that embassy officials summarized for aides to Mrs. Clinton at the State Department's headquarters.

It wasn't breaking news. It wasn't about wars or nuclear weapons or terrorists—the type of news regularly covered in microscopic detail. It was merely an unvarnished assessment about what might constitute the most serious threat to the world's largest food company, which regularly assesses factors that might significantly harm its ability to make a profit or feed the world.

Analysts in the Pentagon and national security agencies pay close attention to rising threats to natural resources and commodities like food because they can lead to armed conflict. Global grain shortages, for instance, have historically caused massive food price increases that led directly to uprisings. The Arab Spring revolution in several countries was sparked by such a spike in food prices, which was a direct result of just such a global grain shortage. That food price surge pushed 150 million people into poverty in 2008, the World Bank has reported.[5] And the United Nations warned that "world grain reserves are so dangerously low that severe weather in the United States or other food-exporting countries could trigger a major hunger crisis next year," *The Guardian* reported in 2012.[6] In addition, rising food prices threatened disaster and unrest.

So while few read about the Nestlé briefing outlined in the 2009 diplomatic cable in the news coverage of WikiLeaks, the threats outlined in that cable are hugely consequential. They speak to the very nature of conflicts over natural resources that national security officials have begun to focus on seriously in the past decade. In short, it *is* news—just not the type of news that American consumers hear with any sort of regularity because it's complicated, messy, and intricate.

"Nestlé, the world's largest food company, worries more about the planet's growing fresh water shortage than the current financial crisis, which it sees as only a bump in the road in the firm's long-term development," U.S. embassy officials summarized in that secret diplomatic cable.[7]

What Nestlé officials described in great deal is extraordinary . . . and alarming. They had assessed the world's entire water supply—its absolute, upper limit on what could be withdrawn sustainably from all of the known, available supplies of fresh water. In 2008, we were using half of the available global fresh water—roughly 6,000 cubic kilometers of the 12,500 cubic kilometers available on an annual basis.

But, Nestlé officials said, a third of the world's population will start to experience fresh water scarcity by 2025—which is now less than a decade away. And the combined pressures of rapid population growth, meat consumption, and demands on other agricultural practices could exhaust all of our available fresh water within our lifetimes. Fresh water scarcity could become dire within a decade and be "potentially catastrophic" by 2050.

"Nestlé estimates the upper limit on sustainable global fresh water withdrawals to be 12,500 cubic kilometers per year, with 2008 use running at about 6,000 cubic kilometers. However, rising population, growing meat consumption, and new biofuel demands are predicted to absorb the surplus entirely by 2050," the authors of the diplomatic cable wrote.

On present trends, Nestlé thinks one-third of the world's population will be affected by fresh water scarcity by 2025, with the situation only becoming more dire thereafter and potentially catastrophic by 2050. Problems will be severest in the Middle East, northern India, northern China, and the western United States.

The company thinks averting a global crisis will require four strategies: (1) creation of a virtual market for water; (2) elimination of subsidies and compulsory preferences for biofuels; (3) universal adoption of more productive, water-efficient genetically modified plants; (4) and the liberalization of global agricultural trade.

At the core of all of this is Earth's rapidly changing climate system, which is driving extraordinary searches for fresh water in certain parts of the earth experiencing intense and regularly occurring multiyear droughts. Earth's changing climate system is also wreaking havoc in other parts of the world, melting glaciers that feed freshwater river systems.

The rate of carbon dioxide increases being deposited invisibly in the black sky above us has been roughly doubling every decade or so since the dawn of the Industrial Revolution. It shows no sign of abating yet, despite pledges from hundreds of world and business leaders to slow it down and eventually reverse the effects of rising carbon emissions.

However, companies that speak green can sometimes act the opposite. In early 2017, two car companies were in the news for emissions-cheating software. "A day after Volkswagen [pleaded] guilty to criminal conspiracy and agreed to pay a $4.3 billion fine in federal court for equipping over 600,000 vehicles" with such software, the U.S. Environmental Protection Agency accused Fiat Chrysler of outfitting "more than 100 diesel cars with secret software allowing them to obscure the amount of pollutants they emit and 'cheat' vehicle emissions tests," reports *Years of Living Dangerously*.[8]

Hundreds of peer-reviewed scientific assessments, policy studies, and summaries from relevant federal science agencies and global conservation NGOs have drawn direct, explicit connections between Earth's changing climate system and the threat of fresh water scarcity on a regional basis.

"In many areas, climate change is likely to increase water demand while shrinking water supplies. This shifting balance would challenge water managers to simultaneously meet the needs of growing communities, sensitive ecosystems, farmers, ranchers, energy producers, and manufacturers," the U.S. EPA has said.[9]

"Fresh water is crucial to human society—not just for drinking, but also for farming, washing and many other activities," the Grantham Institute in London has said. "It is expected to become increasingly scarce in the future, and this is partly due to climate change."[10] The institute further explained that only 2 percent of our water on Earth is fresh. As climate change causes polar ice to melt into the sea, this treasured source of fresh water becomes sea, or salty, water.

"Many of the water systems that keep ecosystems thriving and feed a growing human population have become stressed. Rivers, lakes and aquifers are drying up or becoming too polluted to use. More than half the world's wetlands have disappeared. Agriculture consumes more water than any other source and wastes much of that through inefficiencies. Climate change is altering patterns of weather and water around the world, causing shortages and droughts in some areas and floods in others," the World Wildlife Federation has said.[11]

"The evidence of change is all around us, and not just in the form of the rapidly rising thermometers that are the most commonly cited evidence of the greenhouse effect," says Dr. Peter Gleick, one of the world's leading authorities on water resources.

We are seeing larger and more severe algal blooms in our lakes and along our coasts. The California drought has been deeper and more sustained than it would otherwise have been, as have other droughts around the world. Extreme precipitation events are becoming more frequent and damaging. And pressures, including political pressures, on already scarce water resources are growing.

Even without climate change, the world has plenty of severe water challenges. Seven hundred million people lack access to safe and affordable drinking water. More than two billion lack adequate sanitation. Droughts and floods are already the most damaging extreme natural events for society. Under investment and poor management of water systems leads to preventable contamination events and water-related diseases, especially for poorer, disadvantaged communities. Tensions and violence over shared water resources are growing. Climate change is an added burden on all of these water problems.[12]

U.S. military and national security experts have begun to trace internal civil unrest and armed revolutions to fights over water and food shortages in parts of the Middle East and Africa in the past ten years. WikiLeaks diplomatic cables showed that government officials in both Yemen and Syria were beginning to warn American officials that water scarcity was creating chaos and would almost certainly lead to armed conflict.[13]

That's precisely what happened. In 2011, clashes erupted in Sana'a, Yemen's capital, "shattering a truce between loyalist troops and dissident tribesmen, as security forces shot dead seven protesters," *The Sydney Morning Herald* reported.[14] In Taiz, south of Sana'a, witnesses said that "security forces were opening fire on anyone who tried to gather"—even peaceful protesters.

But for all of Yemen's political troubles in the news, "the headlines do not reveal the part that water plays in this crisis," *The Guardian* said. "13 million Yemenis—50% of the population—struggle daily to find or buy enough clean water to drink or grow food. As a result 14.7 million Yemenis currently depend on humanitarian aid."

In some rural areas, collecting water can take four to five hours a day. Girls spend all their days fetching water and never get an education, says Christopher Ward, author of *The Water Crisis in Yemen: Managing Extreme Water Scarcity in the Middle East*. He adds, "Lack of access to fresh water is notorious for being the biggest cause of malnutrition, morbidity, and mortality in rural areas." Each year, fourteen thousand Yemeni children under the age of five die of lack of nutrition and diarrhea.[15]

But Nestlé's internal assessment, revealed in WikiLeaks, was the first to deeply analyze the fresh water threat to a leading global business enterprise—and, by extension, to billions of people who literally cannot live without access to fresh water. Because the company has business operations in nearly every country in the world, the scope of the threats it faces are global in nature by definition. Thus, Nestlé could draw a clear, unvarnished picture for the State Department officials.[16]

Nestlé's chief economist and director of international relations, Herbert Oberhaensli, led the embassy briefing. Nestlé is a sober, risk-averse company that operates on a long-term basis by recognizing and responding to serious threats before they manifest publicly. The company's ability to prosper in 2008—even as the rest of the world was reeling from a severe global recession that nearly toppled a number of first-world economies—is an excellent example.

While the global economy was in shambles in 2008, Nestlé had one of its best years ever. The company's global sales actually rose 3 percent that year, despite serious inflationary pressure on the Swiss franc. Nestlé's profits during a serious global recession were up 60 percent. "Oberhaensli explained that Nestlé operates on a long-term basis, neither expanding quickly in boom times nor shrinking in recessions," the embassy officials wrote.

Nestlé's reaction to the 2008/2009 global recession was to scale back investment projects, focus on core operations, and hedge against long-term threats like global fresh water scarcity. Its prudence and planning in the face of serious threats explains its unlikely success even in the midst of a global recession. The fact that it viewed fresh water scarcity as an even greater threat than the potential collapse of the global financial economy speaks volumes.

Oberhaensli said long-serving managers at Nestlé were not especially worried by the prospect of a global financial depression, even as its younger managers were somewhat unnerved by the threat of a financial meltdown. "There is a generational split among Nestlé managers in the degree of nervousness with which they view the world situation," the embassy noted. "The younger generation had never seen anything like the current financial turmoil and is quite agitated. In contrast, long-serving Nestlé managers are calmly philosophical, pointing out that the firm has always continued to grow in good times and bad, since there is always solid demand for the company's food products."

The company also noted that consumers in emerging economies were demanding food products that are common in the United States and Europe—a trend that puts an enormous strain on fresh water supplies. "Economy dietary patterns are increasingly converging with those of the U.S. and Europe. Nestlé has seen a huge increase in the demand for milk products such as ice cream and yogurt in China, amounting to a cultural shift." Nestlé co-markets General Mills breakfast cereals outside the United

States. Here, too, diets and lifestyles are converging, which in turn puts added demand on water supplies. "More and more Asians, Europeans, and Latin Americans are starting the day with a bowl of cereal, according to Nestlé," the cable said.

For these reasons, Nestlé told the State Department officials that it views the world economy, and overall food production globally, almost exclusively in terms of a "water economy." And it is this "water economy" that now threatens to destabilize our way of life and the livelihood of hundreds of millions of people within a decade. Water scarcity, Nestlé believes, is what will wreak havoc. "Its management is convinced that growing shortages of fresh water, rather than land, will become the Achilles heel of global agricultural development. [It] is the most dangerous near-term threat to the planet's well-being," the cable said.

A single calorie of meat requires ten times as much water to produce as a similar calorie of other food crops, such as grains, vegetables, and fruits. As the middle class in countries like China and India demand more meat, Earth's water resources "will be dangerously squeezed."

Nestlé then got quite specific with its explanation of the challenge. "Nestlé reckons that the earth's maximum sustainable fresh water withdrawals are about 12,500 cubic kilometers per year," the embassy officials wrote.

In 2008, global fresh water withdrawals reached 6,000 cubic kilometers, or almost half of the potentially available supply. This was sufficient to provide an average 2,500 calories per day to the world's 6.7 billion people, with little per capita meat consumption.

The company estimates that continuing population growth and modest further increases in per capita meat consumption will push annual water withdrawals to 10,000–11,000 cubic kilometers by 2050. This amount will be sufficient to provide the then 9 billion people 2,500 calories a day with somewhat higher per capita meat consumption than today. However, it will require a level of fresh water withdrawals only 15% shy of the sustainable planetary maximum.

But this is where the picture grows dark, and potentially catastrophic. The average American diet isn't 2,500 calories a day—it's closer to 3,600 calories a day (and growing) due to substantial meat consumption. If China and India demand to eat and live precisely as Americans do, then the world will run out of enough fresh water to feed even our existing population on the planet.

"Oberhaensli said that . . . if the whole world were to move to [America's dietary] standard, global fresh water resources would be exhausted at a population level of 6 billion, which the world reached in the year 2000," the cable said. "There is not nearly enough fresh water available to provide this standard to a global population expected to exceed 9 billion by mid-century."

Nestlé has studied water use in crop-growing in most of the countries where they operate and concluded that the main reason crops are grown in so many dry regions (which is now further complicated in the face of multiyear droughts exacerbated by climate change) has more to do with subsidies that mask the true cost of water than anything else.

"Growing a calorie of food crops in a hot dry climate such as California requires much more water than elsewhere," the embassy cable said. "Current water withdrawals in some areas of the world are already unsustainable. The water table is dropping precipitously in the Western US and in northern India. In both areas, users are withdrawing more water than can be replenished and rising salinity is reducing the productivity of plants."

Nestlé said that a few countries in parts of the world facing multiyear droughts are now aware of the serious threats from water scarcity and are taking economic, diplomatic, and even military steps to assure that food and water is available for their citizens.

"Oberhaensli said that Saudi Arabia has recognized that its water is even more valuable than oil and it has decided to stop using irreplaceable fossil water to irrigate food crop fields in the desert in favor of importing the cereals it needs," the embassy cable said.

Not surprisingly, Nestlé sees a global business opportunity in the face of this imminent fresh water crisis. It will take a combination of strategies to avert the crisis, the company believes. At the top of its list of mitigation strategies is the "creation of a virtual market for water, so that the scarce commodity can be traded across borders and so that its price reflects its actual scarcity value."

For instance, an advanced water-pricing system exists in some desert areas of Oman, where water has been scarce and valued for a long time. Water from wells is free for people to drink and for mosques. Beyond that, however, fresh water is divided daily into property shares, which are sold off by minutes of flowing water. A family with more property rights might have access to twenty minutes of a daily flow of water—while others might only have access to ten minutes.

Nestlé also supports genetically modified crops that place far less demand on water supplies, a stance that most scientists would support but has been an extraordinarily difficult sell in the face of protests from anti-GMO activist groups. The company would also like to see world and business leaders liberalize trade agreements to allow food to be grown in parts of the world that aren't dry and experiencing growing drought conditions. Such a concept, however, flies in the face of growing economic nationalism even in countries like the United States that once championed free trade.

For this reason, Nestlé doesn't talk publicly about these sorts of solutions to the looming water scarcity crisis. "Oberhaensli said that . . . these strategies are politically controversial. Sensitive to its public image, Nestlé has maintained a low profile in discussing solutions and tries not to preach," the embassy officials wrote. "For example, in India it does not tell farmers supplying milk to Nestlé factories how to treat their milk cows. However, it pays a premium for low bacteria and high protein counts. [It] offers to assist farmers to introduce advanced production methods. Nestlé has taken GMOs out of certain products due to market pressure, especially in Europe. However, internally the company remains convinced that they represent the future of agriculture."

Despite its reluctance to speak publicly about the grave water scarcity threat it sees in the near future, embassy officials concluded in the secret diplomatic cable that it believed Nestlé honestly wanted to warn government officials around the planet. "Nestlé's concern for the global water supply is genuine," they wrote. Water scarcity could lead to enormous problems in the near future, and "potentially push the price of food out of reach for the poor in some developing nations."

15

Saudi Arabia

The water started to go away in Saudi Arabia fifteen years ago. It was a mystery—in a place where mysteries had confounded religious and world leaders for thousands of years. But this was different. This particular mystery threatened the kingdom and the very fabric of its way of life.

Water in the desert of Saudi Arabia is life itself. Without it, nothing else works. No amount of oil wealth can manufacture crops to feed the people without water.

The kingdom has always had a distinctly difficult problem. While its land holds some of the largest oil reserves in the world, it also holds one of the world's smallest reserves of water. There is not a single lake or river in Saudi Arabia.

For thousands of years, the people of the kingdom have drawn water from wells or the occasional oasis. Infrequent rainfalls replenish shallow aquifers 150 feet or so underground. Wells dug ten times as deep as those shallow aquifers tap into water reserves that are not renewed by rainfall. Once water from those deeper reserves is used up, they're essentially gone forever.

Fifteen years ago, those wells and aquifers started running dry. Saudi Arabia's leaders had placed a very large bet that they could grow certain types of crops in a land without much water to begin with. They started to offer wheat farmers five times as much for a bushel of wheat as wheat farmers in other parts of the world. Much as they had drilled for oil to create massive wealth, Saudi Arabia's rulers tapped those deeper aquifers for water to irrigate crops that sprang forth from the desert.

That bet began to fail at a colossal scale when a combination of factors—from global climate change that accelerated droughts in regions of the world like Saudi Arabia to over-farming in a land where water was such a precious commodity—clearly started to overwhelm the land.

"I remember flowing springs when I was a boy in the Eastern Province. Now all of these have dried up and you have to dig," Abdulrahman al-Qusaibi, Saudi Arabia's water minister at the time, told *The New York Times* in 2003.[1]

Saudi farmers kept digging deeper and deeper wells to find water needed to irrigate their fields of wheat. Some gave up and abandoned their fields. The desert—which historically makes up 90 percent of the Saudi kingdom—quickly recaptured the land in places where deep wells pulled up minerals along with the water, essentially rendering the irrigation useless for crop-growing.

At the time, during the heady days of vast Saudi oil wealth that seemed to have no end in sight, the rulers were extravagant in their use of water. Muhammad H. al-Qunaibet, a hydrologist who advised the Saudi rulers at the time, estimated for *The Times* that Saudi Arabia was using more than six trillion gallons of water a year for agriculture. Only a third of that water was replaced through rainfall. Where did the rest of it go? That was the mystery. It simply vanished.

Deepening the mystery was a somewhat existential question for the rulers of the Saudi kingdom, who have historically taken it upon themselves to provide for every need of the citizens of the country. Just as oil industry analysts and Saudi leaders alike have tried to estimate how much of the Middle Eastern country's vast oil reserves are left belowground, they were also now forced to ask a similar question about its water reserves: How much water was left in the aquifer reserves below the desert?

A survey in 1984—twenty years earlier—estimated that its water reserves intermingled with its fossil fuel reserves were in the neighborhood of 132 trillion gallons. At least half of that water reserve had likely disappeared by the start of 2000. Today? Most of it is probably gone for good.

Eventually, an enterprising Saudi banker put the pieces of the puzzle together and solved at least part of the mystery. In a 2004 report called "Camels Don't Fly, Deserts Don't Bloom," Elie Elhadj showed that the culprit was, in fact, wealthy farmers who had been draining the deeper water reserve aquifers unchecked for several decades—an effort that had turned Saudi Arabia into one of the world's largest wheat exporters but had also caused the water to essentially go away for good.

"A combination of money and water could make even a desert bloom, until either the money or the water runs out," Elhadj wrote.[2]

More recent surveys have shown this rather conclusively. The water is gone in Saudi Arabia. A 2011 research project in the historic town of Tayma, which has thrived for thousands of years above an oasis, found that "most wells [were] exsiccated"—which means they were no longer providing water. An oasis that had literally provided the water of life for more than two thousand years had been drained in a matter of decades.[3]

All of this left the Saudis with an almost intractable problem. Water was already a precious commodity in the Middle East, which holds less than 1 percent of the world's fresh water. Every country in the Middle East pays close attention to political talk by its neighbors about the one natural resource that is both simultaneously imperiled by Earth's changing climate and desperately needed for life itself.

When Turkey's leaders speculate about damming the Tigris and Euphrates Rivers, Iraq and Syria get nervous. Ethiopia and Sudan both have access to the Nile River, which worries Egypt's political leaders. Israel has argued with neighbors about water for decades. Any action the Saudis take to solve its water problem is heavily scrutinized by its neighbors. The kingdom's rulers are constantly accused by neighboring countries of draining water reserves in shared aquifers.

At first, the Saudis tried to conserve the rapidly depleting water reserves in their deep aquifers, which mainly contain nonrenewable water. They formed a Ministry of Water. They stopped paying wheat farmers a small fortune to grow wheat, much of which wasn't actually needed to feed people in the kingdom. They encouraged people to cut consumption of water by about half by using water-saving taps and showerheads. They got smarter about water reclamation and irrigation practices and financed costly deep underground irrigation systems.[4]

But the damage had clearly been done already. The deeper water reserves were essentially gone a decade ago. The Saudis were being forced to live on the much smaller amounts of "renewable" water that was replenished by rainfall—which itself was now being challenged by the uncertain chaos from a changing climate system that rewards or punishes certain parts of the planet with alternately abundant or scant rainfall patterns.

Several years ago, one of Saudi Arabia's sheikhs started a company that was designed to seize water from the headwaters of the Nile in Ethiopia. The plan was to capture the water and bring it back to Saudi Arabia in order to irrigate the wheat fields. The plan provoked an outcry from leaders in every country that relies on the Nile for its own fresh water.

The effort didn't fare well. For instance, local militants invaded the Saudi company's camp in the spring of 2012 and killed five workers there. As a direct result, government soldiers retaliated by raiding local villages nearby, torturing men and raping women. A leading NGO, Human Rights Watch, interviewed some of the people fleeing the conflict to South Sudan nearby. The refugees said the conflict began as retaliation against the Saudi company for seizing its land and water.[5]

These sorts of conflicts, combined with the disappearance of water in Saudi Arabia itself, eventually convinced the rulers that they needed to find another route to grow food for its people—even if it meant going to the ends of the earth to grow crops and bring it back to the country.

In 2015, the Saudis essentially threw in the towel. The rulers announced that the 2016 wheat harvest would be its last. The era of exporting wheat to feed others was over. Because their drinking water was essentially coming from desalinated water from the ocean—which is far too expensive a process for crop irrigation purposes—Saudi's leaders had come to realize that the only way they could feed its thirty million people was to go elsewhere.[6]

The Saudis clearly had to do something. So they came to America.

They weren't alone. China, which is now experiencing many of the same problems the Saudis have experienced with the depletion of permanent water reserves in deep aquifers far belowground, has also come to America for food sources that rely on vast amounts of fresh water. For example, the U.S., combined with Brazil and Argentina, is responsible for nearly 90 percent of world soybean exports, and about 60 percent of those soybeans go to China.[7]

China essentially purchased America's largest pork producer, Smithfield Foods, in 2013 through a company that had the full financial and political backing of its government. A quarter of all pigs raised in America—a process that consumes vast amounts of water to grow the grains that serve as feed for the pigs raised for slaughter—are now part of China's efforts to feed its own people.

The Saudis, though, went far beyond China's efforts. They bought vast tracts of land in Arizona in order to grow alfalfa, which they then export back to the kingdom.

Saudi Arabia's largest dairy producer, Almarai, bought fifteen square miles of farmland in Arizona's desert in order to grow alfalfa—a crop that's so water-intensive that it takes as much as four times more irrigation than wheat.[8] This $47.5 million transaction is an example of the Saudis' efforts to continuously secure its supply of high-quality hay and ensure the country's dairy business, as well as conserving the nation's resources.[9]

Water for Almarai's irrigation efforts come from the same source of fresh water—the Colorado River—that provides drinking water for cities like Los Angeles and Las Vegas. The Colorado River reservoirs have been experiencing all-time lows, creating a volatile local political situation.

What drove the Saudis to Arizona? A secret, classified diplomatic cable released by WikiLeaks—an international, nonprofit organization that publishes sensitive, secret information for public awareness—in 2010 provides the answer. In 2008, then king Abdullah ordered Saudi food companies to find and purchase foreign land with access to fresh water. The king offered to subsidize their operations.

The head of the American embassy in Riyadh wrote in the confidential cable that the effort was needed as a way of "maintaining political stability in the kingdom."[10]

That's how the Saudis came to Arizona and bought fifteen square miles of desert in order to pull fresh water from the Colorado River in an effort to grow water-intensive alfalfa for export back to the kingdom. The Saudi rulers didn't want to risk a revolution, like the Arab Spring, fueled by lack of food and water that might compound other problems, such as political or climate instability.

"Food security continues to be a concern in Saudi Arabia," the U.S. embassy chief wrote in that same 2008 diplomatic cable summary back to State Department headquarters at the end of the administration of President George W. Bush. "The Saudi Arabian government is seeking alternative sources for grain importation. Rising costs of food have caused considerable concern among lower income groups, both Saudis and guest workers. If inflation continues unabated, it could undermine political stability in the Kingdom."

U.S. embassy officials were certain of Saudi intentions because it had come directly from a meeting with the kingdom's top trade and commerce officials at the time: Taha Alshareef, assistant director general for foreign trade, and Ahmed Al Sadhan, general manager of the National Office for Industrial Strategies at the Ministry of Commerce.

Even then, in 2008, it was obvious to Saudi's senior career government officials that the grand experiment to pull fresh water from the deep aquifers to grow wheat was failing. "Al Sadhan stated the need for Saudi Arabia to seek alternative grain sources in response to rising global food prices," the U.S. embassy chief wrote. "Saudi Arabia has 1.76% arable land, and water scarcity makes it impossible to sustain the current levels of grain production in the Kingdom."

While Saudi Arabia has historically struggled with food security concerns, recent developments at the time had driven the Saudis' rulers overseas in a frantic search for farmland where they could grow crops to feed their people.

"Historically, all grains have been imported into Saudi Arabia, apart from wheat, which was grown in the Qassim region using fossil water from the aquifer. In the 1980s and 90s, the government employed massive subsidies, the result of which Saudi Arabia was even briefly a wheat exporter until water depletion drove them to import all grains," the embassy chief wrote.

Al Sadhan said that the Saudi government was conducting feasibility studies to look at investing in farms in third-world and developing countries—as well as in developed countries like the United States. The plan was to "pay the operating costs of the farms and provide management; in exchange for local land, labor, and a portion of the grain produced (particularly wheat and rice)," the embassy chief wrote. "He also said the countries in which the farms operated could use the remainder of the grain produced to ameliorate their own rising food costs. Alshareef noted that private Saudi companies have had success with similar ventures in the past."

The order to buy farmland in other countries came directly from then king Abdullah—with orders to keep the effort secretive "for fear that target countries might inflate the cost of farm-land in anticipation of investment."

The U.S. embassy chief noted that the Saudi government's interest in investing in farmland outside its borders "appears genuine for the primary purpose of addressing the fundamental food security issue, though the driver behind this new initiative is the sudden price inflation of basic commodities." He concluded that "there is the risk of civil unrest if food prices become too high, particularly from the lower and middle class Saudis, as well as the foreign worker population. Food price inflation could pose a direct threat to the social contract between the House of Saud and the bulk of the population. Therefore these programs seem aimed less directly at the economy and more at the goal of maintaining political stability in the Kingdom."

By the time the Saudis arrived in America to buy up vast stretches of the Arizona desert, American national security officials had begun to recognize the potential for armed conflict over access to natural resources like water. President Barack Obama's director of national intelligence, James Clapper, delivered a stark speech in 2014 that stated precisely this.

Food insecurity and water scarcity are central elements of the "most diverse array of threats and challenges as I've seen in my 50-plus years in the intel business," Clapper said in that 2014 speech. "As time goes on, we'll be confronting issues I call 'basics' resources—food, water, energy, and disease—more and more as an intelligence community."[11]

By the time that Almarai, Saudi Arabia's largest dairy company, had purchased the Arizona land in order to grow hay for export back to the kingdom, the word had spread. The Saudi king's directive was no longer a secret. Almarai officials said publicly that they were following the king's orders, which had essentially created a "virtual water" enterprise. It was actually cheaper to use Arizona water to grow and ship hay than it would be to bring water in to irrigate Saudi farmland.

None of this sits well with Arizona residents living close to Almarai's operations. They complain that its hay operation is depleting the state's aquifers—much like the Saudis depleted aquifers in its own country for decades. More than three hundred local residents packed a community center in La Paz County in January 2016 to listen to the director of Arizona's water department explain how long the state's desert aquifer might (or might not) last. It was a disquieting meeting.

There were five sheriff's deputies standing guard at the meeting to make sure that it didn't turn ugly. A county supervisor, Holly Irwin, said that the state's water department had requested that extra law enforcement show up for the meeting. "Water can be a very angry issue," she said. "With people's wells drying up, it becomes very personal."[12]

The state's water department director, Thomas Buschatzke, defended the Saudi investment because it provides local jobs and investment. "Arizona is part of the global economy; our agricultural industry generates billions of dollars annually to our state's economy," he told *The Arizona Republic*.[13]

But Buschatzke has also acknowledged that, like the aquifers in Saudi Arabia, he couldn't predict with certainty how much longer the local area's water might last. Arizona, like the rest of the American Southwest, is faced with a changing climate that makes extended droughts in that part of the world more severe.

These changes, combined with the intensity of the Saudi farming practices, could put considerable pressure on the state's water reserves.

Buschatzke promised that the state would study the issue. Following the meeting, his department approved two new wells for Almarai—each of which can pull more than a billion gallons of water a year from the aquifers in the region.

"It's gotten very emotional," said Irwin. "When you see them drilling all over the place, I need to protect the little people."[14]

16

Yemen

Yemen is a small country immediately south of Saudi Arabia, known to Americans largely through an endless parade of stories about terrorism, civil war, political chaos, and pirates.

It was one of the seven predominantly Muslim countries included in President Donald Trump's travel ban (later reduced to six when Iraq was dropped from the list). People trying to enter the United States from Yemen were temporarily banned by the American president because Yemen is known as a breeding ground for terrorists.

Yemen has been overrun by a brutal civil war that has destabilized the central government. In recent years, analysts have begun to predict that it will likely become the next staging ground for pirates who venture out into shipping lanes to seize ships and ransom wealthy passengers—a practice that continues despite routine monitoring by U.S. and European navies. Piracy has a global economy price tag of nearly $18 billion every year, with higher risk for cargo, skyrocketing insurance, and inflated costs for traders and consumers.[1]

Yemen's traditional capital, Sana'a, has been controlled by rebels opposed to the existing government since the start of 2015. As a result, the ruling government has temporarily relocated the capital to the port city of Aden on its southern coastline.

Yemen can seem to outsiders like a brutal, violent place that terrifies both its own citizenry and foreign travelers on a routine basis. It's the poorest country in the Middle East, with few near-term prospects for development while the civil war keeps it in a state of political crisis. With the added continual presence of refugees, increases in the cost of living, and the lack of basic health and social services, Yemen has chronic food shortages, devastating poverty, and heightened gender-based violence.[2]

Saudi Arabia has involved itself in Yemen's politics for years in an effort to keep the peace at its own southern border. An informal political settlement among warring tribes was largely held together through a patronage system that used cash payments from the Saudis to enforce it. The Saudis funneled money through a partner to sheikhs throughout the country.

Much of that process collapsed in 2011, when street protests against widespread poverty and corruption turned violent and eventually erupted into a civil war that still exists today. The street protests were a direct challenge to then president Ali Abdullah Saleh, who was long suspected of running what amounted to a kleptocracy—a government that chiefly seeks status and personal gain at the expense of its people—inside the impoverished country. The results are that natural resources, such as oil and water, that could help build Yemen's future and bring economic and political stability to the country, are nearly depleted.[3]

Saleh was ousted in 2012, a year after the protests erupted in Yemen's version of the Arab Spring. Three years after he was removed from office, a panel of United Nations experts released a political corruption report alleging that Saleh secretly amassed a personal fortune of up to $60 billion during his thirty-three years in power. His fortune included gold, cash, and property held in different names in twenty countries, the UN report said.[4]

Saleh tried to regain power. He'd aligned himself with one of the leading rebel groups (the Houthis, who routinely counter Al Queda insurgents) that captured Yemen's capital of Sana'a, which caused his successor to flee the country. But Saleh was assassinated by a sniper in December 2017.[5]

By any reasonable measure, Yemen appears to be a place of political chaos, terror, corruption, and lawlessness. But that isn't the whole truth— and it hasn't always been that way.

Yemen is the first casualty of what will almost certainly be the "water wars" of the future. What has happened inside Yemen is a harbinger of what will happen elsewhere when water scarcity drives citizens to take desperate measures.

Here's the truth. Even though it is surrounded by oceans on two sides, Yemen is running out of fresh water. This simple fact is at the root of many of its problems. Without water, and then food, virtually every other aspect of a civil society becomes untenable. This is what has happened to Yemen. It is the first example of what a water war looks like.

Outside political and civil society observers tried to warn leaders about the growing threat for years—all to no avail. By the time anyone was paying attention, it was too late.

"Yemen could become the first nation to run out of water," *The Times* of London reported in the fall of 2009—two years before political chaos ripped the country apart in a brutal civil war.[6] Not only is Yemen the poorest country in the Arab world, it is also the most water-scarce in the Arab world as well—a situation that has been made worse by routine, multiyear droughts inflicted on the region by the changing climate.

Even before climate change began to make things worse, Yemen struggled with access to fresh water. Groundwater from rainfall is the main source of fresh water for the country. Water tables have dropped so precipitously in recent years that most parts of the country now have no viable sources of water any longer. In Sana'a, the state capital, the water table has dropped from 30 meters below the surface in the 1970s to 1,200 meters below the surface by 2012. Since few can afford the cost of tap water pumped to their housing, most residents get water from public urban fountains. Sanitation is a chronic problem, with diseases such as malaria rapidly spreading.[7]

People in Yemen have access to levels of fresh water that are less than 10 percent of what most experts consider to be a reasonable level of access to water. The threshold for "water stress"—where a society's demand for water for domestic, industrial, or agricultural uses exceeds the available amount or when poor quality restricts its use—is around 1,700 cubic meters of water per year for an average person. Drop below that level, and the lack of water hampers human health, well-being, and economic development.[8] Today the average person in Yemen has access to only 140 cubic meters of water.

What's worse is that, in a country where agriculture uses about 90 percent of the available water, roughly half of all crops in the country are used to grow a narcotic called *qat* that most Yemenis chew—which means half of the crop production that uses almost all of the water is for a crop that doesn't feed people. Yet approximately four-fifths of Yemen now struggles to find water to drink.[9]

U.S. political experts in the region saw the water crisis coming well before the rest of the world did. Stephen Seche, the U.S. ambassador to Yemen, wrote about it in a secret, confidential cable, captured in the WikiLeaks release in 2010. When water riots erupted in Yemen in 2009 (again, two years before the Arab Spring and the civil war had begun), Seche told CIA and State Department leadership that "water shortages have led desperate people to take desperate measures with equally desperate consequences."[10]

In fact, Seche sent more than one confidential cable back to Washington to warn political leaders that water scarcity was poised to shred civil society in Yemen—a water war that would have ripple effects far beyond Yemen's borders.

"Water remains a socially threatening, yet politically sensitive subject in Yemen; government action has stagnated as water resources continue to decline," the U.S. ambassador wrote in 2009. "The lack of water has resulted in water riots in the governorates of Aden, Lahj, and Abyan."

Seche said that the water scarcity crisis had created other serious problems as well. Dengue fever outbreaks were becoming common. People were starting to hoard water in cities. Despite pressure from the United States and others, no one in Yemen's government was acting on the emerging threat. "There is no one clear partner in the Ministry of Water and Environment and its associated agencies," Seche added.

Clearly Yemen's leaders knew there was a grave problem by 2009. They just didn't do anything about it, perhaps because there was very little they could do once the water tables had dropped to levels never seen before. Yemen's minister of water and the environment at the time, Abdulrahman Al-Eryani, called water scarcity "insidious" and the "biggest threat to social stability in Yemen," the U.S. ambassador wrote.

Again, remember that this was in 2009—a full two years before the Arab Spring took the world by surprise and appeared to erupt from nowhere. But the Yemen uprising and civil war didn't spring from nowhere. Leaders, including U.S. leaders, knew instability was just over the horizon and that it was being driven in large part by a lack of fresh water.

"According [to] Eryani, there are more conflicts over water than land in Yemen. Indeed, water protests occurred in the governorates of Aden, Lahj, and Abyan at the end of August, as people gathered together to demand water and accountability from officials," Seche wrote. Eryani, Yemen's water minister, described the water riots as a "sign of the future" and predicted (quite accurately) that the growing conflict between urban and rural areas over access to water would lead to violence.

Yet despite the fact that both Yemen's leaders and important political observers like the United States knew all too well what was about to happen inside the country in the face of the water riots, no one acted. There were environmental laws on the books, but the Yemeni police, military, and courts didn't enforce them.

"Despite the [Yemeni government] having a comprehensive environmental law and a National Water Strategy, enforcement of the law and implementation of policy remains a challenge," Seche wrote. That's a polite way to phrase it. The truth was, virtually all of the hundreds of rigs that drilled for water in the country were completely unregulated.

"A large part of the problem is that Yemen has over 900 rigs for drilling, largely unregulated by [Yemen's government]," Seche explained. "Drilling rigs, in fact, are subsidized and exempt from custom duties. While the . . . water law is one of the best in the region, according to Eryani, it cannot be implemented without controlling the rigs."

Left unsaid here is that the water rigs were unregulated because they were part of an elaborate scheme to enrich local, regional, and national leaders of warring tribal factions. The government could have nationalized the rigs in order to provide clean, fresh water to the Yemeni people, which seemed like a logical action in the face of such a crisis, but this option was never even on the table.

"An alternative to regulating the rigs is to nationalize them, which would cost $100–150 million," Seche revealed. "Because of illegal rigs, 14 out of 16 aquifers are depleted in Yemen. Eryani said that the Sa'ada aquifer is in the worst condition and the lack of water is one of the causes of the conflict, although no one is talking about it. Governorates particularly affected by water shortages include Sana'a, Aden, Abyan, Lahj, Taiz, and Sa'ada."

Here, in one terrifying paragraph, is the real root of Yemen's horrific problems—the sole issue that has destabilized that entire region and created a situation where terror could breed and travel to the rest of the world. When "14 out of 16 aquifers" are depleted in the country, that means there is no water for people to live. Without water, there are no crops to grow food. And when the political leaders hoard tens of billions of dollars of gold and property around the world, there is nothing left to provide basic necessities for an entire country. This, in a nutshell, is Yemen's plight today.

Saudi Arabia, which struggles with water scarcity as well just to the north, has created a "virtual water" network in order to provide food and water for its people. But Yemen doesn't have Saudi Arabia's wealth—and what little wealth it has accumulated has largely fled the country into secret bank accounts all over the world.

The American public sees only the military conflict there—not the origins of that conflict. When refugees flee from a country like Yemen, it is for the simple reason that they can't find the basic stuff of life there, and they are caught in the wake of the resulting turbulence. Violence and civil war are what people see—not the daily grind of searching for water that is no longer there.

Refugees flee places like Yemen for all sorts of reasons—some of which might also be to export terrorism to the United States and elsewhere. But they mostly flee because they can't find water to drink or food to eat. Those who remain behind fight each other for water and food. This is the future in places like Yemen, whether we acknowledge it or not.

"The water scarcity situation is only projected to get worse as the population increases and agricultural use of water, particularly in the production of qat, expands," Yemen's U.S. ambassador wrote in that 2009 confidential cable. "Yet the effects of water scarcity will leave the rich and powerful largely unaffected."

Even though it was against government regulations, Seche said one wealthy Yemeni businessman simply drilled "his own personal backyard well to a depth of more than 800 meters." Another wealthy Yemeni businessman joked about his rivalry with other wealthy neighbors over competing water wells—both of which drain the same water basin. Neither well offered water to other people outside their family compounds.

"These examples illustrate how the rich always have a creative way of getting water, which not only is unavailable to the poor, but also cuts into the un-replenishable resources," Seche wrote.

When there is no water, even in the largest cities, people grow desperate. They start hoarding water and storing it in their homes in case they can no longer find it. That in turn creates an unsanitary breeding ground for public health epidemics. "Water shortages have led desperate people to take desperate measures with equally desperate consequences. Residents of Taiz are hoarding water, keeping it everywhere in their homes in open containers, providing breeding grounds for mosquitoes carrying dengue fever," Seche explained.

This in turn puts more pressure on an already unstable infrastructure. "Given a sharp increase in severe cases, which must be hospitalized, the Yemeni health system may face capacity issues in treating affected individuals," he wrote.

Meanwhile, even in the face of water riots, food shortages, and public health outbreaks that all pointed to an impending political crisis, the government chose not to act. "Yemen continues to suffer from bad public policy in dealing with the intertwined relationship between water scarcity, agriculture, and public health. The recent water riots in Aden, Lahj, and Abyan and the outbreak of dengue fever in Taiz could be a sign of the future," Seche wrote presciently. Desperate, with no path forward, the people of Yemen "may be relegated to rallying in the streets due to the lack of service."

In the two years preceding the Arab Spring and the political collapse in Yemen in 2011, senior American and European government officials repeatedly tried to warn Yemeni leaders of the impending crisis. In a September 2009 meeting between a State Department deputy assistant secretary, Janet Sanderson, top European diplomats, and senior officials running Yemen's government, they tried to turn the national conversation away from a focus on counterterrorism and security and toward pressing economic and social challenges.

Yemen's senior government officials were receptive to the approach, but they clearly had no ability to influence Saleh, Yemen's then president, to do anything about those challenges. As a result, Yemen's ministries were never able to follow through with a plan to address the underlying problems that were undoubtedly leading to political chaos.

In the meeting, according to a second confidential diplomatic cable from the 2010 WikiLeaks release, "Foreign Minister Abubakir al-Qirbi pressed for a 'strategic dialogue' between the United States and Yemen in order to ensure that the relationship is not dominated by security and counterterrorism issues," U.S. embassy officials wrote in that 2009 cable.[11]

"Economic advisors presented an ambitious plan for achieving their top 10 priorities for economic reform. Environment and Water Minister Adulrahman [sic] al-Eryani urged that Yemen's water crisis, increasingly a driver of conflict and instability, be a major issue on the bilateral agenda, and he asked for political, rather than financial, support to put it there," the cable added, summarizing the meeting between top U.S., European, and Yemen government officials.

"European Ambassadors grappled with how to press Saleh for political and economic reforms, recommending high-level U.S. engagement with Saudi Arabia, and advising U.S. officials to be blunt and 'brutally honest' in their conversations with President Saleh," the diplomatic cable concluded.

"With respect to economic development and addressing the water crisis, Yemeni advisors and officials have formulated thoughtful and realistic reform proposals that will require political (specifically presidential) will in order to have any hope of being implemented."

As we know now, in the wake of both the Arab Spring and the political revolution that destabilized Yemen in 2011, Saleh never heeded the warnings and allowed the country to fall into political chaos and civil war that shows no signs of ending anytime soon. Yemen's ministers knew the crisis was coming. It was patently obvious to them.

In that meeting in the fall of 2009, they told the U.S. and European officials that civil war and political chaos was coming. "[About] 70 percent of unofficial roadblocks stood up by angry citizens are due to water shortages, which are increasingly a cause of violent conflict," the same cable stated. "He [Yemen's water minister] noted that small riots take place nearly every day in neighborhoods in the Old City of Sana'a because of lack of water, and he predicted that the capital could run out of water as soon as next year."

But Yemen's ministers were powerless to do anything—despite the obvious warning signs. "A major obstacle to doing so is that fact that the rig owners are powerful individuals (army officers, sheikhs, members of the president's family, and certain government ministers) who are 'untouchable' by the law," the embassy officials wrote in that cable back to Washington.

Yemen's ministers asked the United States to put pressure on neighboring countries—especially wealthy Saudi Arabia—to intervene before it was too late. "First and foremost, they said, is Saudi Arabia, which plays a critical role in Yemen due to the considerable financial support it provides to both the Saleh regime and hundreds of Yemeni sheikhs on its payroll," the cable said. "KSA [the Kingdom of Saudi Arabia] reportedly has given [Saleh] $300 million in recent months, to prosecute its war against the Houthis and attend to other pressing needs."

Think about that sum from the Saudis—$300 million over the course of just a few months "to prosecute its war against the Houthis" along with other pressing needs. According to the UN report six years later, it's almost certain that most of that money eventually made its way to secret bank accounts outside the country. Some of it was directed to the military. It's almost certain that almost none of it went to what was actually creating the crisis—the lack of water and food for the basic needs of the Yemeni people.

What's more, every senior government official in that meeting knew that the reforms they knew needed to occur in Yemen were, in fact, never going to occur. "Even if [Saudi Arabia] could be convinced to demand more reform from Saleh in return for its support, if unnerved by instability in Yemen, [the Saudis] would likely break ranks and infuse Yemen with cash, without reform strings attached," the cable noted. "The [U.S. and European] ambassadors agreed that threatening to cut off development aid is not an effective lever for demanding political reform. According to the German ambassador, 'Saleh doesn't care if we give $80 million or $200 million in development aid. What he wants is political support against the Houthis and the Southern Movement.'"

Their dire predictions were exactly what came to pass. Yemen erupted in street riots and political chaos in 2011. Civil war descended. Saudi Arabia reacted almost identically to what these government officials had warned privately two years earlier. But the crisis all began with water—or, to be precise, the lack of water throughout the country—and the instability this issue created for people.

An entire country had essentially run out of water. As such, Yemen is a case study for what the world will face as other countries suffer similarly in the coming years and decades. The twin forces of Earth's changing climate system and overuse of fragile water supplies and aquifers will play out in other countries.

Yemen was only the first. There will be others, sooner than most realize.

17

Syria

When people have no food, they lose hope. They rebel against the leaders of their government. In some cases, they become revolutionaries.

When severe droughts hit countries for years on end, farmers give up and move away from their family farms. Grain price increases drive higher food prices, which in turn triggers unease and even revolt.

Analysts have long described how this deadly combination—drought, grain shortages, and then a surge in prices that put food staples beyond the reach of millions of people—sparked the Arab Spring revolution. Food prices pushed 130 million to 155 million into poverty in 2008, the year before the Arab Spring, according to the World Bank estimates.[1] "Prices of main food crops such as wheat and maize are now close to those that sparked riots in 25 countries in 2008," reported *The Guardian*.[2] In the same article, Lester Brown, president of the Earth Policy Institute in Washington, says, "Food shortages undermined earlier civilizations. We are on the same path. Each country is now fending for itself. The world is living one year to the next." He also warns, "The situation we are in is not temporary. . . . We are beginning a new chapter. We will see food unrest in many more places."

Global grain shortages are appearing with more regularity in regions whipsawed by Earth's changing climate system. Some parts of the world—including California's Central Valley, which is responsible for more than a third of the world's food supply—are seeing extended droughts that last for several years at a time.

Syria saw precisely this from 2006 to 2010. A severe drought—the kind of prolonged and extreme event that scientists have long said will become more and more common—had a crippling effect on hundreds of thousands of farmers who simply couldn't survive a drought that lasted for five years.

The extended drought couldn't have come at a worse time for Syria's farmers and the people of Syria who relied on those farms for their food. The country was already beginning to collapse under the weight of hundreds of thousands of refugees fleeing from conflict in Iraq and elsewhere. When the drought hit, it was the match that lit the fuse.

The world now looks at Syria and sees only a brutal civil war, with horrific casualties in all directions and a war that seems to have no end. People rarely take the time to examine the root causes—the beginning of that conflict.

They're hiding in plain sight.

Just as diplomats and government officials saw the water riots in Yemen leading to political chaos and a civil war there years before it actually erupted, the origins of the internal conflict in Syria likewise were evident to a few before Syrian president Bashar al-Assad began to prosecute his deadly military war to maintain power.[3]

At the very end of George W. Bush's presidency, career State Department officials sent a confidential cable back to Washington, D.C., warning that a drought (which was in its third year at the time) was poised to disrupt the fragile Syrian government.

The warning came in the form of an urgent request on November 8, 2008, to the U.S. government from Abdullah Bin Yehia, who was Syria's representative to the United Nations' Food and Agriculture Organization (FAO) at the time. Yehia said Syria desperately needed American foreign aid to help small-hold farmers in northeast Syria who had run out of options.

"Yehia proposes to use money from the appeal to provide seed and technical assistance to 15,000 small-holding farmers in northeast Syria in an effort to preserve the social and economic fabric of this rural, agricultural community," Yehia said, according to a summary of the request from U.S. embassy officials in Syria.

"If [these] efforts fail, Yehia predicts mass migration from the northeast, which could act as a multiplier on social and economic pressures already at play and undermine stability in Syria," the embassy officials wrote in the secret diplomatic cable that was released by WikiLeaks in 2010, two years later.

The FAO's request at the time was relatively modest, but the implications behind it had grave implications—nearly all of which came to pass in the next two years.

The UN's office that coordinates foreign assistance for humanitarian efforts—which is usually reserved for efforts in the wake of extreme weather events like typhoons or military conflicts that create refugee camps in neighboring countries—had begun an appeal in the fall of 2008 to raise enough funds to keep Syria's farmers in business.

The UN Office for the Coordination of Humanitarian Affairs was asking for just $20 million to help "one million people impacted by what the UN describes as the country's worst drought in four decades," the diplomatic cable said.

The foreign aid was needed to pay for a variety of projects, but the most urgent need was to "restore food production and safeguard agricultural livelihoods [and] provide emergency food assistance to the victims of the drought," U.S. embassy officials wrote.[4]

Despite the fact that world leaders like then president Bush and Russian president Vladimir Putin were pouring tens of millions of military dollars into the region to prosecute counterterrorism efforts against Al Queda and the emergence of the Islamic State in the region, not a single country had stepped up in Syria in a meaningful way to help the farmers.

Syria's farmers were on their own three years into the drought. No one thought to help them in a time of such dire need so they could stay on their farms and continue producing needed crops. Instead, their only option seemed to be to try moving to the cities, with the slim hope of finding jobs there. However, in the cities, there were no jobs and, thus, no food. Yet no one in the political realm seemed to recognize that, where there is no food or water or jobs, revolutions and civil wars are born.

"There does not appear to have been much movement on the part of donor countries to fund this appeal [for just $20 million to help Syrian farmers plant crops] thus far," Syria's FAO representative said.

Italy had chipped in $700,000. China committed $500,000. Greece—yes, the same Greece that was on the verge of its own economic collapse at the time—had pledged $250,000. These were paltry amounts considering that the United States and Russia were spending hundreds of millions of dollars on military actions nearby.

The Syrian FAO representative warned in the fall of 2008 that a "perfect storm" was coming. No one listened.

Yehia told a U.S. Department of Agriculture official who had visited the region to assess the threat facing Syria's farmers that "a confluence of drought conditions with other economic and social pressures . . . could undermine stability in Syria."

Yehia was desperate. No one was providing any money, at all, to help farmers in Syria who were on the edge of giving up their family farms for an uncertain future in Syria's major cities, so he had narrowed his appeal to try to help the small-hold farmers in Al-Hasakah, the hardest-hit region of the country in northeast Syria.

Yehia was being strategic with his narrow request. He reasoned that if he could get some assistance to that hardest-hit part of Syria, other aid might follow for the rest of the country's desperate farmers. Al-Hasakah shares a northern border with Turkey and a southern border with Iraq. Mosul, the focus of the Islamic State's efforts, is just an hour's drive from Al-Hasakah.

"Because the UN appeal has, thus far, not been entirely successful, Yehia has had to prioritize aid recipients," the embassy official wrote cryptically in the secret cable.

The UN boiled down its request so narrowly, in fact, that only a very small subset of Syria's farmers would qualify for the small amount of aid that was still not forthcoming. To qualify, a farmer from Al-Hasakah had to be the head of the household, hold no more than two hectares of farmland, list crops as the sole source of revenue off the land, hold no more than eight cows on their farm—and report "zero" as a crop yield in 2008 and "no seed stock" needed to replant for 2009.

In the face of what was about to happen in Syria, this seems patently absurd. Syria was already struggling with a million refugees from nearby Iraq, which had still not recovered from a devastating war. The seeds of the Islamic State revolution were abundant, in all directions. The plight of the Syrian farmers was so obviously the final straw for a country that was teetering on the edge that's it's hard to understand, in retrospect, how no one saw it coming.

"Yehia does not believe that [Assad] will allow any Syrian citizen to starve," the embassy officials wrote. The Syrian government had stockpiled enough wheat to provide subsidized bread at least through the coming winter. "However, Yehia told us that the Syrian Minister of Agriculture, at a July meeting with UN officials, stated publicly that economic and social fallout from the drought was 'beyond our capacity as a country to deal with,'" the embassy officials noted in the cable.

This is precisely what Yehia was trying to head off. The Syrian FAO representative was begging for a very small amount—only $20 million to help Syria's farmers survive the extended drought and replant for the 2009 season. Without it, Syria's farmers would have no choice but to give up—with devastating consequences.

"What the UN is trying to combat through this appeal, Yehia says, is the potential for 'social destruction' that would accompany erosion of the agricultural industry in rural Syria. This social destruction would lead to political instability, Yehia told us," the cable noted.

"He fears that up to 15,000 rural farmers [in Al-Hasakah, near Mosul] will fail to plant crops for the 2009 growing season, either because they have no residual seed stock from the disastrous 2008 crop, or because they are hesitant to plant seed (bought on credit) for fear of a repeat of the 2007–2008 cold snap that killed seedlings last year," it concluded.

The warning was clear and direct. With a small amount of aid, Syria's farmers would likely stay on their family farms and replant for the 2009 season. Without it, they would give up, abandon their farms, and move to the country's large urban areas, where there were no jobs and where the seeds of a growing rebellion against Assad were already growing.

"Without direct FAO assistance, Yehia predicts that most of these 15,000 small-holding farmers would be forced to depart Al Hasakah province to seek work in larger cities in western Syria (Damascus and Aleppo, primarily)," the cable stated. "Approximately 100,000 dependents—women, children and the elderly or infirm—would be left behind to live in poverty, he said. Children would be likely to be pulled from school, he warned, in order to seek a source of income for families left behind."

Worse, he warned that the farmers—who had no skill beyond farming— would overwhelm Syria's major cities. "The migration of 15,000 unskilled laborers would add to the social and economic pressures presently at play in major Syrian cities," the cable said. "A system already burdened by a large Iraqi refugee population may not be able to absorb another influx of displaced persons, Yehia explained, particularly at this time of rising costs, growing dissatisfaction of the middle class, and a perceived weakening of the social fabric and security structures that Syrians have come to expect and—in some case—rely on."

None of it worked. No one helped.

The United States turned away FAO's straightforward request for aid to Syria's farmers. "Given the generous funding the U.S. currently provides to the Iraqi refugee community in Syria . . . we question whether limited [U.S. government] resources should be directed toward this appeal at this time," the State Department official concluded in the face of the FAO request. Western efforts in the region continued to focus on massive amounts of military aid to deal with the aftermath of the Iraq war and the emerging Islamic State threat.

As a direct result of a severe drought that would last another two years, a million Syrian farmers, herders, and their families left their small-hold family farms and moved to Damascus and Aleppo—just as Yehia had predicted. Few found jobs. Many joined the rebellion against Assad in the major urban areas of Syria.

These unskilled former farmers migrated to major cities along with another million refugees fleeing the ravages of war and the Islamic State. All of these moves amounted to a cauldron of uncertainty, much of which might have been prevented with a relatively small amount of agricultural aid from other countries that had the financial means to do so.

But the greater point is that what happened in Syria is a glimpse of the future awaiting us if we don't see the clear warning signs.

Syria's experience is the future in parts of the world where the earth's changing climate system is creating severe droughts, earlier or later rainy seasons that sometimes include crippling deluges, and occasionally extreme warm days that can reach as high as 130 degrees Fahrenheit in parts of the Middle East.

Extreme or uncertain climate system changes make it difficult for farmers to grow crops—in Syria or anywhere else on the planet. Where there is no food, there is often no hope and revolutions begin.

Syria and other countries in the Middle East presently face this uncertainty going forward. *The International Journal of Climatology* published a study in 2013 that concluded "consistent warming trends since the middle of the 20th century across the region" were leading to "increasing frequencies of warm nights, fewer cool days and cool nights."[5]

Syria's experience tells us something clearly. Droughts and extreme weather created fear and uncertainty. No one responded to or headed off the seeds of political chaos emerging as farmers left their fields, despite warnings from experts. When government officials finally acknowledged the problem, it was too late. The damage had been done.

"By 2010, roughly one million Syrian farmers, herders and their families were forced off the land into already overpopulated and underserved cities," *New York Times* columnist Thomas Friedman has written. "These climate refugees were crowded together with one million Iraqi war refugees. The Assad regime failed to effectively help any of them, so when the Arab awakenings erupted in Tunisia and Egypt, Syrian democrats followed suit and quickly found many willing recruits from all those dislocated by the drought."[6]

A number of academic, peer-reviewed studies have now assessed what really happened in Syria that led to its civil war. All of them point to what should have been obvious—that the extended drought forced farmers from their fields to the cities, leading to food shortages and unrest in the cities where there were no jobs for the farmers who had no job skills other than farming.[7]

"Drawing one of the strongest links yet between global warming and human conflict, researchers said that an extreme drought in Syria between 2006 and 2009 was most likely due to climate change, and that the drought was a factor in the violent uprising that began there in 2011," *The New York Times* reported in 2015.[8]

"The drought was the worst in the country in modern times, and in a study published [in the] *Proceedings of the National Academy of Science*, the scientists laid the blame for it on a century-long trend toward warmer and drier conditions in the Eastern Mediterranean, rather than on natural climate variability."

In the midst of the clear warnings by both scientists and experts, it's obvious that we are seeing a new class of refugees—climate refugees—who are leaving their homes in the face of a growing threat, but in yet another strange and weirdly surreal circumstance, no one recognizes this rather obvious fact in the twenty-first century.

There is no such thing as an official climate refugee, anywhere. Even the United Nations does not recognize such a category. There are refugees of war and terrorism, but for millions who are forced to flee from places where there is no food or water? They are nameless, with no official status as refugees.

"This amorphous, global population of refugees does not have any international legal protection or agency upholding their basic human rights and helping to keep them safe," an Al Jazeera investigation found in 2015.[9]

The reason, Al Jazeera found, is that they aren't covered by a refugee convention that was formed in the shadow of World War Two—a convention that hasn't been updated since.

"Yes, there is a protection gap involving climate change refugees, but we don't call them climate refugees for the reason that they are not covered by the 1951 [Refugee] Convention," Marine Franck, a climate change officer at the UN's refugee agency, UNHCR, told Al Jazeera.

The refugee treaty extends "only to people who have a well-founded fear of being persecuted because of race, religion, nationality, or membership of a social group or political opinion, and are unable, or unwilling to seek protection from their home countries," Al Jazeera found.

What of a million Syrian farmers and their families who flee from their farms in the face of a five-year drought, which scientists connect to Earth's changing climate system? They don't count.

"This means that the estimated 200,000 Bangladeshis who become homeless each year due to river erosion cannot easily appeal for resettlement in another country," Al Jazeera reported. "It also means that the residents of the small islands of Kiribati, Nauru and Tuvalu, where one in 10 has migrated within the past decade, can't be classified as refugees, even though those who remain are trapped in worsening environmental conditions."

War happens for many reasons and creates all manner of refugees. We know the root causes in Syria. We saw them coming . . . even if we still can't properly name them.

18

Jordan

Hundreds of thousands of refugees are trapped between countries in camps in the dry, dusty plains of Jordan. Until, or unless, world leaders recognize an independent Palestinian state someday, these refugees inside Jordan literally have no home that they can truly call their own.

More than two million registered Palestinian refugees live in Jordan. Most of them have full citizenship—but not all. There are ten recognized Palestine camps in Jordan, which hold nearly four hundred thousand refugees without a country of their own.

Jordan hosts more Palestinian refugees than any other country in the world. It also shelters hundreds of thousands of refugees fleeing from the conflict in nearby Syria. These refugees, especially, suffer from remorseless, abject poverty. Their legal status is precarious. Health care and education are difficult propositions.[1]

However, these terrible problems—tenuous legal status, awful health-care delivery services, substandard schools—may be the least of their worries.

Jordan was already one of the driest countries on the planet before a decade of incredibly dry rainy seasons at the start of the twenty-first century made a bad situation considerably worse. Jordan has witnessed the "driest season in decades," the water ministry said in a statement two years ago.

The total rainfall in Jordan in 2014 was less than a third of the country's long-term average annual precipitation—in a country that was already incredibly arid and dry and one of the world's most water-scarce countries. In 2014, the country's dams held just 43 percent of their total capacity.

Even in good years, Jordan's annual water supply is far below the recognized water "poverty line." Its annual per capita water supply of 145 cubic meters is far below the international water poverty line of 500 cubic meters per year.[2]

Jordan lies squarely within the subtropical band above Earth's equator that scientists have predicted for years would become progressively drier and subjected to prolonged drought as the climate system changes.

The rainy season that usually provides just enough rain to support crops in Jordan is now coming later and later each year. The changes are small each year, but over time, it adds up to a profound shift in just the space of a decade.

A study by Abdul Halim Abu-Hazeem, the former head of Jordan's Meteorological Department, found that rainfall in Jordan fell overall by 8 percent from 2001 to 2010 and that the onset of the rainy season kept slipping year after year.

Crops like wheat and barley that are fed by rainfall during this season have been adversely affected, experts say. Two years ago, farmers in Jordan began to notice a growing trend where wheat was standing in fields in February each year at heights far below where they were in previous years.

The trend had become obvious to them over the course of two decades, they told government officials in a special report on Jordan's growing agricultural sector crisis by IRIN, a nonprofit closely aligned with the humanitarian arm of the United Nations.[3]

Jordan's rainy season should be under way by the end of September or early October. In 2014, the first rainfall in Jordan didn't come until November 1—a trend now noticeable to every farmer in the country, said Adel Abdul Ghani, a professor of agriculture studies at Mu'tah University in Karak Governorate.

Jordan's farmers rely on rainfall that falls heaviest during the months of December and January. Wheat in February in Jordan used to stand two feet tall or more. In 2014, farmers noticed it was standing a quarter of that (less than half a foot) at the same time of the year.

"Repeat dry years due to changing rainy seasons and a shortage of rainfall observed since 1994" discouraged many farmers from planting their fields in 2014, as they were "no longer profitable," Ghani said.

Farmers and agricultural experts alike say the dry weather in Jordan is only a continuation of poor weather over the past few decades that has left them struggling. All of this is consistent with what scientists have long described as inevitable.

Years ago, Jordan was actually able to export wheat. Today, like Saudi Arabia, that is no longer an option, as a combination of increasing dryness and shrinking water tables have made wheat production a difficult proposition.

The president of Jordan's Farmers' Union, Audeh Rawashdeh, said that wheat exports are nothing more than a fond memory for Jordan's farmers at this point. "We used to export wheat and grains, but now what we produce is only enough to meet the country's needs for [a few] days."

Each year, Jordan is forced to import more wheat to feed people. The country imported 95 percent of its wheat needs in 2010, and 98 percent in 2011, according to Jordan's Department of Public Statistics.

Unlike the United States—where Republican leaders in Congress have delayed acknowledging the growing impacts of climate change on agriculture, coastlines, snowpacks, and other elements of Earth's ecosystems because it threatens the long-term business models of large energy companies—Jordan's agricultural scientists decided years ago that they could no longer wait for the inevitable. They began an effort to develop crop seeds that were resistant to drought and water scarcity.

Jordan's national center for agricultural research has been studying "resilient wheat" for nearly a decade. The effort is specifically designed to adapt to what is so obviously a changing climate system in their part of the world. Jordan's scientists, out of clear necessity, are trying to adapt and cope with the increasing dryness by developing a strain of wheat that can deal with less water and a delayed growing season.

"By surveying a variety of wheat grown locally in Jordan, we realized that an adaptive trait is early flowering and early maturity, which helps plants escape dry weather," said Nassab Rawashdeh, a researcher on climate change and biodiversity at Jordan's national agricultural center.

The new variety of wheat is working. "It has succeeded [in southern Jordan] under extremely dry conditions," Rawashdeh said. But in a sign of how scientific advances progress in fits and starts, some farmers are leery of the strange, new variant of wheat. Therefore, the new drought-resistant wheat has not been adopted countrywide.

Olives, a hugely important cash crop for Jordan's farmers, have similarly been hurt by the gradually uneven and chaotic rainy season, the IRIN special report found.

"Farmers used to wait for the first rainfall to harvest their olive trees, but due to the delays nowadays many harvest them before," said Mohammad J. Alatoom, who runs the environment and climate program for the UN's development office in Amman.

When a cold snap in 2014 destroyed more than half of the olive trees in northern Jordan, more farmers simply gave up. "It is awful, when it is your only source of income," a farmer, Nabeela Meghdadi, told IRIN, adding that she'd decided to look for other sources of income because farming had "proved unreliable."

Rawashdeh, the national agricultural researcher, said that farmers were being whipsawed by slow changes in water scarcity and arid conditions at one end and extreme weather events at the other that are starting to identify a climate signal.

The cold snap in 2014 that destroyed so much of Jordan's olive crop is "a form of extreme weather event resulting from climate change," he told IRIN for the special report. "This is a result of climate change impact. This is severe weather which we are not used to dealing with."

An emerging water scarcity crisis, however, might prove even more debilitating for Jordan, creating a cascading impact on the fragile state of affairs that has existed in the Middle East for decades.

A 2014 report by Mercy Corps, one of the largest humanitarian groups in the world that works closely with the ten main refugee camps as well as the Syrian refugees, said that water scarcity will soon place enormous pressure on the refugee populations that consistently present such a vexing political problem in the region.

"In the current crisis over water, Jordan faces a perfect storm of pressures. Refugee demands layer over long-standing challenges of scant supply, unsustainable management, and out-of-date infrastructure," Mercy Corps said in the report.[4] "But refugees also bring new challenges. These range from poor conservation habits to an overwhelming volume of human waste that, improperly treated, threatens to pollute groundwater."

With hundreds of thousands of refugees—and many more fleeing the Islamic State and Syrian civil war every day—water issues in Jordan go well beyond the fact that it is the third-most water-scarce country in the world to begin with. The random infrastructure that has grown up around the refugee camp population in Jordan is chaotic, disorganized, and unwieldy.

In fact, vast amounts of fresh water never even make it to taps in Jordan's households each year because it is literally dumped into the sand in the country's broken infrastructure system, Mercy Corps found.

"Jordan, one of the world's driest countries, is dumping much of its water into the sand. Aging infrastructure is the culprit. Of all the water Jordan pumps, billions of liters never reach a family's tap. Instead it gushes out of broken pipes," the NGO found. "The amount of water lost nation-wide could satisfy the needs of 2.6 million people—more than a third of Jordan's current population. It is a tragedy of waste."

Compounding the volatile political and humanitarian situation are studies showing that Jordan could simply run out of fresh water within a generation through a combination of factors that include overpumping, a completely dysfunctional infrastructure system, and increasing pressure from a changing climate.

"Water scarcity is slated to worsen, threatening health and stability in one of the Arab world's most durable states," Mercy Corps said.

Given the chaos in Syria that is driving hundreds of thousands of new refugees their way, the situation will only become worse.

"Few doubt that after Syria's bloody civil war pushed 600,000 people across the border, refugee pressures have sped up the clock," the humanitarian relief agency said in the report. "Groundwater depletion has accelerated; water tables are dropping precipitously. And as water levels decline, salinity rises, rendering what remains less and less drinkable."

The surging Syrian refugee population is threatening to destabilize the country and the region overall. "In some communities, water demand has quadrupled," the Mercy Corps report said. "Supply has not kept pace. Weeks might elapse before a drop comes out of the tap. Increasingly, families don't have enough water to drink."

In the past, Jordan's leaders have been able to deal with increasing dryness and drought conditions by flooding the camps and cities with water from groundwater supplies. Those days of indulgence are long gone. Jordan has run out of time, Mercy Corps concluded.

"By over-exploiting groundwater, the Government of Jordan was able to meet [daily water needs]. But no more," the report stated. "Those communities hardest hit by refugees have seen the average supply drop below [the water poverty line]. At that level, sanitation standards decline, diseases rise, subsistence crops wither, and children go thirsty."

Just as we've seen in Yemen (where water riots led directly to a bloody civil war) and Syria (where drought drove farmers to major cities, prompting food riots and civil unrest), the situation in Jordan has reached a crisis point, Mercy Corps said.

"Jordanian patience is wearing thin. Just as troubling is Jordanians' growing frustration with their government, [which] has been a generous host to beleaguered Syrians," Mercy Corps concluded. "But the institutions responsible for providing water have been unable to meet rising demand. Anger over water scarcity is increasingly directed at Amman. In some cases, tensions have already ignited."

Other studies have mirrored the Mercy Corps study. One, by a national research center in Switzerland, said that Jordan's water crisis had reached a critical juncture.

"Jordan faces a deepening water crisis, exacerbated by climate change, regional conflict, immigration, and poor governance. Its people are among the most water-deprived worldwide," according to the 2013 working paper from the Swiss National Centres for Competence in Research.[5]

"If a path to water sustainability is to be found, a nationwide, coordinated approach to parallel political and water reforms combined with imaginative regional diplomacy over shared and new supplies will be indispensable," the Swiss center concluded in the working paper.

There is some slim hope on the horizon—at least for drinking water. After decades of fighting over water issues, Jordan signed a controversial water-sharing agreement with Israeli and Palestinian authorities that is supposed to go into effect sometime in 2018.

The agreement includes the construction of a desalination plant at the Port of Aqaba in Jordan that will be shared by Jordan and Israel and pump brine to revive the Dead Sea. A pipeline and canal system will run to and from the plant.

"There is no other way Jordan can address water scarcity, given the increasing population and challenges brought by climate change," Nabeel Zoubi, a Jordanian project manager, told *The Guardian*.[6]

The problem—which is often the case for such endeavors in the region—is financing. The World Bank has supported the project in concept, but financing for such an endeavor is complicated, and large. It may never materialize.

For its part, Mercy Corps believes big, institutional changes must occur in Jordan before it's too late. A single desalination plant at Aqaba won't solve the problem.

Mercy Corps recommended a number of key measures needed to keep Jordan's water crisis from erupting into conflict. Its most important recommendation was to actually fix the infrastructure problem. Just six engineers are responsible for making sure millions of Jordanians and refugees have water to drink or grow their crops.

"Government actors are under-resourced and undermanned; their capacity badly needs an upgrade. Jordan's front-line water utility is emblematic," Mercy Corps said in one of its three principal recommendations to deal with the water crisis.[7]

"Responsible for an area larger than Hawaii and with a Jordanian and Syrian population of millions, it is paralyzed by lean operating budgets and a staff of six engineers," the NGO said. "These shortcomings must be addressed. Investing in infrastructure makes little sense if new projects are handed over to agencies that have neither the resources nor expertise to run and maintain them."

But if history in the region is a guide, the recommendations will simply be ignored until it's too late. Water and food riots could very well lead to civil war and armed conflict—as they have in Yemen and Syria.

19

Somalia

Somalia has heard of wars, and rumors of wars, for a generation. Over the past twenty years, foreign powers great and small have launched mostly unsuccessful proxy wars to seize control of the impoverished nation along the eastern coast of Africa. Somalia is the public face of the Horn of Africa, which juts hundreds of miles into the Arabian Sea and Indian Ocean.

The sad, unseemly tales of military misadventure in Somalia over the past two decades are on ghastly display in several mass releases of formerly secret diplomatic cables in both the United States and Saudi Arabia.

In 2006, for instance, the United States foolishly convinced Ethiopia to invade.[1] Several years later—with the United States, the Saudis, and others whispering in its ears through diplomatic channels[2,3]—Kenya launched its secret "Jubaland Initiative," designed to carve out a separate state in southern Somalia.[4] Other foreign powers like Qatar have been accused of playing both sides, alternately funding both rebel groups and provisional Somalian governments.[5]

These military actions are often described by political leaders as a necessary means to defeat the fundamentalist jihadist militant group Al-Shabaab—an offshoot of the Islamic Courts Union that was defeated by Somalia's Transitional Federal Government in 2006—rather than a naked power grab. Al-Shabaab aligned itself with Al Queda in 2012.

The most recent release—a huge cache of cables from Saudi diplomats, which are still being translated from Arabic to English—show yet again how foreign powers are more than willing to wage costly, adventurous wars to seize control of various parts of Somalia. The Saudis, likely interested in Somalia's untapped oil reserves, were eager to join in the war games designed to carve up Somalia into as many as four separate states, these cables show.[6]

But while the United States, Russia, Saudi Arabia, Kenya, Ethiopia, and every other foreign power that covets Somalia's oil reserves and verdant coastline and ports were secretly conspiring to carve up the country for their own economic or military purposes, a human and environmental disaster of unimaginable proportions was quietly unfolding in Somalia.

It is a disaster that has no real parallel in the long, sordid history of colonial, imperialistic adventures in Africa. It is also, as we've seen in other neighboring countries like Yemen, at the very center of the terrorism and military conflicts that have surrounded the Horn of Africa for the better part of twenty years.

On March 10, 2017, United Nations' Humanitarian Affairs and Emergency Relief Coordinator Stephen O'Brien delivered a shocking report to the UN National Security Council on the mind-numbing humanitarian crisis facing Somalia, Yemen, South Sudan, and Nigeria.[7]

Unless the major countries that form the National Security Council intervened with an unprecedented level of humanitarian aid—on the order of billions of dollars of new humanitarian aid in the space of just a few months—more than twenty million people in these countries would die from starvation and famine.

"We are facing [the] largest humanitarian crisis since [the] creation of [the] United Nations," said O'Brien, a reserved British diplomat who became the head of the UN's humanitarian and relief agency in the spring of 2015. "20M+ face starvation, famine," he wrote on social media. Without global efforts, he said, "people will starve to death."[8]

O'Brien said that the crisis in these four countries was unlike anything the world had seen since World War Two and would require an enormous, immediate level of humanitarian aid at unprecedented levels in order to keep those twenty million from dying.

The desperate plea from the UN's humanitarian agency came at a most unfortunate time. It landed just a matter of weeks after newly elected President Trump had released his first budget guidelines to Congress—a budget that called for a massive cut in U.S. contributions to the United Nations as well as a substantial reduction in foreign aid overall.

Trump and the GOP Congress were turning inward and closing America's borders at a time when ecological, economic, and military disasters were wreaking havoc with impoverished countries on the other side of the world.[9]

The Republican Party has long hoped to scale back the U.S. commitment to the United Nations and reduce the foreign aid budget as well. President Trump's first budget was built to respond to those pledges by GOP leaders to pull America back from the rest of the world on humanitarian campaigns, while simultaneously increasing America's capacities as a military superpower.

The stark humanitarian UN report—outlining the imminent deaths of millions unless nations respond with an unprecedented level of humanitarian aid—was released within this political context. The gulf between these opposing worldviews—one that demands a global humanitarian response and another that holds fast to an "America First" framework built on military strength—is the widest it has been since the United Nations was formed in the shadow of the Second World War.

Of all the stories contained in the O'Brien report to the UN National Security Council, Somalia's was by far the worst—a looming tragedy that is nearly impossible to comprehend. More than half of Somalia's entire population will die of starvation without help, O'Brien told the Security Council.

"In Somalia, more than half the population—6.2 million people—need humanitarian and protection assistance, including 2.9 million who are at risk of famine," he said.[10] O'Brien said that more than $4 billion was needed by July as immediate, emergency assistance. Without it, half of Somalia's population would likely die.

For people living in America—even in the poorest parts of America—this level of suffering and tragedy doesn't seem quite real. How is it possible that half of the population of an entire country might starve to death in a matter of months? How is this actually possible? How had Somalia arrived at such a precarious time?

Here's how.

While the United States and other large foreign powers were obsessively focused on destroying Al-Shabaab, hunting pirates off the coastline of the Horn of Africa, launching proxy wars to seize control of coastlines and oil reserves, and plotting ways in which to section off Somalia into as much as four different political states, extended droughts and water scarcity were doing the real damage. Earth's changing climate system was quietly and insidiously making life intolerable in Somalia.

Before he delivered his shocking report to the UN Security Council on March 10, 2017, O'Brien and several of his staff extensively toured Somalia and the other three nations in order to assess the magnitude of the environmental and humanitarian disaster facing them. Their assessment reports served as the basis of the overall recommendation to the Security Council for immediate help.

What O'Brien found in Somalia was horrific—but hardly surprising to agricultural experts and earth scientists who have been tracking the unfolding ecological disaster in that part of the world.

"Close to 1 million children under the age of 5 will be acutely malnourished this year," O'Brien said. "In the last two months alone, nearly 160,000 people have been displaced due to severe drought conditions, adding to the already 1.1 million people who live in appalling conditions around the country."[11]

That word—*acutely*—is a polite word for imminent death by starvation. It means that hundreds of thousands of children are about to die without immediate food assistance that somehow manages to make its way past marauding Al-Shabaab militants and other militant groups across the Somalian landscape.

But it is the ecological disaster O'Brien's staff discovered that should wake people up and cause them to rethink our approaches in this part of the world. This is the real disaster in Somalia, and it will only get worse.

"What I saw and heard during my visit to Somalia was distressing—women and children walk for weeks in search of food and water," O'Brien said. "They have lost their livestock, water sources have dried up and they have nothing left to survive on."

What is occurring in Somalia right now has been fifteen years or more in the making. The "desertification of Somalia" has been under way for years. Experts have warned since at least 2001 that Somalia was mortgaging its long-term future for short-term gains. The changing climate system, which leads to extended drought conditions in parts of the world like the Horn of Africa, has made things exponentially worse.

"[Somalia is] struggling to survive in an often harsh environment, made worse by the ever-growing danger of environmental degradation and desertification," the United Nation's Development Program said sixteen years ago in a 2001 ecology report.[12] "Ironically, some efforts to earn a living—clearing land for agriculture, producing charcoal, overgrazing herds on shrinking pastureland, selling timber for construction—contribute to the problem, [and] are jeopardizing Somalis' ability to eke out a living from the land in the future."

Somalia, like Yemen across the gulf, was trapped between the viselike grip of economic deprivation and a merciless ecological cycle. "Somalia is caught in a vicious cycle where poverty and desertification are intertwined," UN official Randolph Kent said in that 2001 report. "You can't address one problem without addressing the other."

That 2001 UN report outlined any number of practices that were contributing to the problem that would eventually erupt in 2017 into an unprecedented humanitarian crisis.

"Several practices are contributing to desertification," the 2001 UN report said.

Clearing land along riverbanks to create more area for agriculture is causing rivers to change course and eroding nutrient-rich soil. Land clearing is now especially intense along the Juba river, as the population of the southern port town of Kismayo swells and the demand for agricultural produce grows.

The sudden rise in Kismayo's population is related to an influx of people engaging in the booming charcoal trade, a profitable but environmentally devastating export business. Three large forested areas comprised mainly of acacia bussei trees are quickly being cut down to feed the charcoal export market. Local estimates indicate that as many as 1 million 25-kilo bags of charcoal, worth approximately $6 per bag, leave Kismayo each month destined for the Gulf states.

Deforestation in parts of the country would also allow the desert to return, the UN predicted. "The clearing of mangrove trees along the coast, in the northwest and parts of southern Somalia, is contributing to sand dune encroachment, which threatens farm land and the network of coastal roads. The trees are cut for timber and also to create more agricultural land," the report concluded. Overgrazing of pasturelands would also contribute to desertification as well, it said.

The United Nations had high hopes fifteen years ago for interventions "aimed at addressing these and other environmental concerns, with a focus on helping to preserve land and water resources. Such measures can halt the spread of desertification and mitigate the impact of drought."

They all failed. Somalia's various governments and foreign powers alike focused, instead, on military campaigns time and time again. Al-Shabaab, pirates, and military invasions were planned and launched . . . while the desert returned and harsh, uncompromising environmental factors made the country virtually inhospitable for human life.

In his 2017 report to the UN Security Council, O'Brien was uncompromising in his assessment of what years of neglect had done to Somalia.[13] The story here is a virtual carbon copy of what occurred in Syria years earlier, which researchers have now traced to the epicenter of the military conflict.

"With everything lost, women, boys, girls and men now move to urban centers," O'Brien said.

As in Syria, there are no jobs or food in either the cities or the large, growing refugee centers. Where there is no food or jobs in these large population centers, rebellion and eventually revolution grows unchallenged.

António Guterres, the newly minted UN secretary general, joined O'Brien and his staff on one of the assessment trips to see firsthand what was occurring in these large cities as people fled the land to urban areas and refugee centers. They visited Baidoa, which became the capital of the South West State of Somalia in 2012.

"With the Secretary-General—his first field mission since he took office—we visited Baidoa. We met with displaced people going through ordeals none of us can imagine. We visited the regional hospital where children and adults are desperately fighting to survive diarrhea, cholera and malnutrition," O'Brien told the council.

In the midst of such suffering, it was easy to ignore the harsh ecological realities while military atrocities dominated the news coverage in the region. Military conflicts and acts of terrorism made it difficult, if not impossible, for humanitarian efforts to operate in the region.

"Again, as if proof was needed, it was clear that between malnutrition and death there is disease," O'Brien said.

Large parts of southern and central Somalia remain under the control or influence of Al-Shabaab and the security situation is volatile. Last year, some 165 violent incidents . . . directly impacted humanitarian work and resulted in 14 deaths of aid workers.

Al-Shabaab, government forces and other militia also continue to block major supply routes to towns in 29 of the 42 districts in southern and central Somalia. This has restricted access to markets, basic commodities and services, and is severely disrupting livelihoods. Blockades and double taxation bar farmers from transporting their grains.

The tragedy unfolding in Somalia in 2017 has echoes from the past. More than 250,000 people died in the 2011 famine that struck Somalia. The scope of the tragedy unfolding in Somalia, however, is exponentially larger—and will require a coordinated response never witnessed before in that part of the world.

"We can avert a famine," O'Brien vowed. "We're ready despite incredible risk and danger. [But] we need the international community . . . to invest in Somalia. [We] need those huge funds now. To be precise we need $4.4 billion by July, and that's a detailed cost, not a negotiating number."

Whether the United States and other first-world nations join with major philanthropic foundations to avert the immediate humanitarian disaster that unfolded in Somalia in 2017 or not—and there was a measured response from various parties that forestalled the worst, for now—the underlying problem that has threatened the country's future and long-term hope for survival will remain long after the aid workers leave.

Somalia, like Yemen, now faces an uncertain ecological, environmental, and agricultural future. Earth's changing climate system is creating chaos on top of disastrous agricultural and political choices. There is no easy answer in Somalia, regardless of humanitarian and military responses.

While it is nearly impossible to grasp the notion that half of a country's population is constantly at imminent risk of dying from starvation and famine, the much larger question for the rest of the world is this: Is Somalia a harbinger for what may yet occur in other parts of the world?

20

Pakistan

Pakistan is one of the fastest-growing nations in the world. It had a population of 170 million in 2011. Five years later, that population crossed the 200 million mark. The country is grappling with the same sorts of growing pains that its neighbor India is experiencing.

But Pakistan has an extraordinary problem looming on the horizon: Water scarcity, which has devastated other countries in the subtropics in the past decade, is now quite real. And a solution to the crisis is not entirely within the country's control.

The Indus River is the primary source of fresh water for most of Pakistan. It's responsible for much of the water that's used in both Pakistani households and industries. Water from the Indus also supports 90 percent of the agricultural sector in Pakistan—a particular problem for a country that, like others in the subtropical regions of the world, is arid and dry to begin with.

The Indus, like the Nile in Egypt, is one of the great rivers of the world, but the river has been so exploited in the past two decades—even as dry conditions grow worse in subtropical regions—that it no longer even flows into the ocean at the Port of Karachi.

The Indus is "dribbling to a meager end. Its once-fertile delta of rice paddies and fisheries has shriveled up," water expert and author Steven Solomon has written in *The New York Times*.[1]

Once a lush ecosystem, the lower Indus and the varied habitat it supports is now threatened in myriad ways.

"Choked off from its water supply, Karachi is plagued by increasingly brazen water thieves and riots over scarcity. Many in the water-stressed delta blame wealthy landowners upstream for taking water out of the river," *National Geographic* reported in a special series on global water issues.[2]

But here's where it gets especially treacherous for Pakistan. Compounding the overuse and changes inflicted on the arid region from Earth's climate system, actions by India to cut off some of the flow of water feeding the Indus has created the potential for serious conflict between the two nations.

The glaciers that feed the Indus originate in India, which has implemented large-scale diversions of the fresh water as it cascades down from those glaciers. India has even bigger plans for diversions. This, not surprisingly, has created considerable tension with Pakistan.

"One of the potentially catastrophic consequences of the region's fragile water balance is the effect on political tensions," *National Geographic* reported.[3]

In India, competition for water has a history of provoking conflict between communities. In Pakistan, water shortages have triggered food and energy crises that ignited riots and protests in some cities. Most troubling, Islamabad's diversions of water to upstream communities with ties to the government are inflaming sectarian loyalties and stoking unrest in the lower downstream region of Sindh.

But the issue also threatens the fragile peace that holds between the nations of India and Pakistan, two nuclear-armed rivals. Water has long been seen as a core strategic interest in the dispute over the Kashmir region, home to the Indus' headwaters. . . . [D]windling river flows will be harder to share as the populations in both countries grow and the per-capita water supply plummets.

A bit of context is necessary here to understand how severe a problem this is right now for Pakistan and how it can become catastrophic in the near future.

The Intergovernmental Panel on Climate Change (IPCC)—a definitive report on the causes and impacts of climate change globally compiled by thousands of scientists every four years—has signaled for nearly a decade now that dry regions of the world in the subtropics will continue to see less and less rainfall.[4] Some of this is already occurring. The Horn of Africa (which includes Somalia, Yemen, and Kenya) falls squarely in the subtropics where decreased rainfall has a severe impact on already dry regions. India and Pakistan do as well. The overall effect of climate change is an intensification of the water cycle that causes more extreme floods and droughts globally. The subtropical regions of the world are ground zero for these impacts.[5]

An IPCC special report on climate change adaptation says that at least a billion people in subtropical regions of the world like Pakistan, India, Yemen, Saudi Arabia, and Somalia will face increasing water scarcity. These subtropical regions will suffer badly from reduced rainfall and increased evaporation, the IPCC concluded.[6] As we've seen in Somalia, these impacts on top of shortsighted industrial agricultural practices and deforestation are even now creating horrific humanitarian challenges.

That's the backdrop for the growing conflict between Pakistan and India over fresh water and the Indus River. Nearly all of agriculture in Pakistan relies on the Indus. So do Pakistani households and industries. If India continues to create large-scale diversions of water that flow into the Indus, the potential for conflict between India and Pakistan will become very real. Verbal jousting will turn into water riots or even armed conflict. The situation is further exacerbated by the fact that both India and Pakistan possess nuclear weapons.

Glacier melting is responsible for roughly half of the water flowing in the Indus, making the situation worse. The health of the Himalayas in the face of Earth's changing climate is a real, and growing, concern.

"Given the rapid melting of the Himalayan glaciers that feed the Indus

River . . . and growing tensions with upriver archenemy India about use of the river's tributaries, it's unlikely that Pakistani food production will long keep pace with the growing population," Steven Solomon wrote in *The New York Times*.[7]

The potential for conflict—including military conflict—between India and Pakistan over the Indus River is one of the reasons that former president Barack Obama and his then secretary of state, Hillary Clinton, added "water" as a diplomatic priority early in the previous administration.

"By 2025, two thirds of the world will live in water-stressed areas. One billion will face outright water scarcity," the Pulitzer Center reported on a briefing from then undersecretary of state Maria Otero. "As rising populations face dwindling resources, the probability of conflict will increase."[8]

In her briefing, Otero also stressed the growing challenge from global warming on countries like Pakistan and India in subtropical regions. Changes in weather patterns will cause some regions to see intensified drought while leaving others drenched in rain.

In the summer of 2010, in fact, Clinton announced a massive foreign aid package built around water security that was primarily designed to benefit Pakistan. The $7.5 billion aid package was built to bolster national storage capacity, irrigation, and safe drinking water in Pakistan.

Beyond the conflict with India, Pakistan has an enormous infrastructure problem on its hands. The country has the world's largest continuous irrigation system, and it is littered with all kinds of problems along its route that have been left unattended for far too long.

"To some extent, these deficiencies have been masked since the 1970s by farmers drilling hundreds of thousands of little tube wells, which now provide half of the country's irrigation," writes water expert Solomon. "But in many of these places the groundwater is running dry and becoming too salty for use. The result is an agricultural crisis of wasted water, inefficient production and incipient crop shortfalls."[9]

Just as we've seen in Yemen—where water riots ripped the country apart and led to a civil war that has destabilized the country in the midst of political chaos—wealthy, politically connected landowners in Pakistan have also been accused of siphoning off far more than their fair share of fresh water in upriver Punjab. There have been water riots over lack of water and electricity in Karachi.

"The future looks grim," Solomon concludes. "Eventually, flows of the Indus are expected to decrease as global warming causes the Himalayan glaciers to retreat, while monsoons will get more intense. Terrifyingly, Pakistan only has the capacity to hold a 30-day reserve storage of water as a buffer against drought."

Moves by India have aggravated an already tense situation. The Indus Waters Treaty between India and Pakistan was hammered out in 1960 to share the Indus River. The treaty is designed to make sure that volumes of fresh water downstream aren't diminished by industrial uses or dams upstream.

When India built a series of hydropower dams where the tributaries feeding the Indus emerge from the Himalayas, it didn't technically violate the treaty, but experts believe that India's efforts to dam up the Indus could ultimately destroy Pakistan's ability to feed its population.

If both countries collaborated on a series of giant, large-scale dams that were built to rotate water use to different regions, tensions could be reduced, but that level of cooperation between nations over water use has never truly succeeded on the scale needed to deal with the looming threat.

The potential repercussions aren't contained to this region. What happens in the region affects the global economy as well. As one example, a third of the world's cotton supply comes from India and Pakistan. More than seven hundred billion gallons of water are pulled from the Indus River every year to grow this cotton.

"Pakistan's entire economy is driven by the textile industry," says Michael Kugelman, a South Asia expert at the Woodrow Wilson International Center for Scholars. "The problem with Pakistan's economy is that most of the major industries use a ton of water—textiles, sugar, wheat—and there's a tremendous amount of water that's not only used, but wasted."[10]

The public anxiety and tensions about the potential conflict over water between Pakistan and India, however, are even more direct when Western diplomatic analysts talk behind closed doors. In a series of secret State Department cables released by WikiLeaks, the analysts don't pull any punches.

"Pakistan claims that India diverted a significant volume of water from the river Chenab to its Baglihar dam, which resulted in a 34 percent drop in Pakistan's water levels," said one secret cable. "After numerous talks with their neighbor nation, Pakistan now plans to take its concern to the World Bank for compensation under the terms of the Indus Waters Treaty."[11]

Pakistan claimed that India was causing rolling blackouts in India by diverting water from the Indus. "The water shortage will likely lower winter crop yields as irrigation levels are lower and has already resulted in increased blackouts due to reduced hydroelectric power generation," said a second cable.[12]

At one point during an especially tense set of negotiations between the two countries, David Mulford, the U.S. ambassador to India during the Bush administration, wrote that Indian dams being planned along the Indus could lead to war.

"Even if India and Pakistan could resolve [existing] projects, there are several more hydroelectric dams planned for Indian Kashmir that might be questioned under the IWT [Indus Water Treaty]," Mulford wrote in another confidential cable.[13] While he didn't believe it would happen, he did acknowledge privately that "Islamabad's worst case scenario [is] that India's dams . . . have the potential to destroy the peace process or even to lead to war."

A decade ago, India and Pakistan seemed content to wage their water wars behind closed doors, in confidential diplomatic cables, or in the halls of third-party arbiters like the World Bank, the United Nations, or the International Monetary Fund. No more. In the fall of 2016, things took an ominous turn.

India and Pakistan have fought three wars over Kashmir. Both lay claim to it and also administer portions of it. At the end of September 2016, India's military staged a surgical strike inside the parts of Kashmir administered by Pakistan in an effort to target seven terrorist camps.

India announced the military strike publicly. Pakistan denied that it even took place (though it did later admit that Pakistani soldiers were killed by Indian soldiers along their disputed border). Making the situation much worse, India's strike inside the parts of Kashmir controlled by Pakistan happened less than two weeks after a Pakistani terrorist group (Jaish-e-Mohammed) killed eighteen Indian soldiers on a military base in a town in India-administered Kashmir.

Both Indian and Pakistani ministers swiftly escalated the political rhetoric, which is typical in instances such as this. Pakistan's defense minister threatened regional nuclear war. An Indian minister countered that Pakistan was a "terrorist state." However, this is just talk. It pales in comparison to an actual threat that was nearly carried out by India in the midst of the back-and-forth rhetoric.

After the Indian soldiers were killed in Kashmir, Vikas Swarup, a spokesman for India's Foreign Ministry, said that India might consider revoking the Indus Waters Treaty. "For any such treaty to work," said Swarup, "it is important for mutual trust and cooperation. It cannot be a one-sided affair." A move by India against the treaty would instantly be seen as a real act of war by Pakistan.

Four days later, India's government met to consider just such a move. While it ultimately decided not to revoke the Indus Waters Treaty—for now—India left open the clear possibility of doing so at a later date. India's prime minister, Narendra Modi, issued a cryptic, yet ominous, threat during the review. "Blood and water cannot flow together," Modi said.

India then immediately suspended meetings that had been planned between treaty commissioners who routinely meet to keep the critical water treaty intact—a move that was designed to send a clear, chilling message to Pakistan's leaders.

Pakistan reacted immediately. A spokesman for the Pakistani prime minister said that revoking the water treaty would be an "act of war."

"If Pakistan's access to water from the Indus Basin were cut off or merely reduced, the implications for the country's water security could be catastrophic," Michael Kugelman wrote in *Foreign Policy* magazine as the events were unfolding in real time. "For this reason, using water as a weapon could inflict more damage on Pakistan than some forms of warfare."[14]

After decades of inattention to irrigation, overuse, and pollution along its portion of the Indus—combined with the overarching threat of climate change—Pakistan is now at a dangerous crossroads.

Pakistan is presently one of the most water-stressed countries in the world, according to the International Monetary Fund. Its per capita annual water availability shrank to the water scarcity threshold in 2016—at a time when the country's water intensity rate (a measure of cubic meters used per unit of GDP) is the highest in the world.[15]

These two things—water scarcity combined with the highest intensity of water use anywhere in the world—are a recipe for catastrophe in Pakistan. If India makes good on its promise to revoke the Indus Waters Treaty, there is simply no telling what Pakistan might be forced to do to protect its people.

"Pakistan's economy is the most water-intensive in the world, and yet it has dangerously low levels of water to work with," Kugelman wrote.[16]

The situation grows direr in light of Pakistan's groundwater tables. NASA satellite data released in 2015 showed that the underwater aquifer in the Indus Basin was the second-most stressed in the world. When surface water supplies disappear, countries turn to aquifers. Pakistan is perilously close to disaster in both areas. More than forty million people in Pakistan already lack access to safe drinking water. A move by India against the treaty would send Pakistan over the edge.

As if all of this were not ominous enough, China is watching the India-Pakistan water wars closely. Beijing has never signed any international water-management treaties. It could, if it wanted, build its own series of dams even farther upstream in the Himalayas and cut off fresh water access to India. The Indus originates in Tibet, and China could conceivably cut off the flow of the Indus River into India.

In Pakistan—like Yemen, or Somalia, or Saudi Arabia—water is life. Even if India continues to honor its commitments under the Indus treaty, it faces a very difficult path forward in a part of the world where politics and ecology are poised to create a deadly mixture.

21

India

It's like clockwork. Water from the Indian Ocean is carried by trade winds in early June to mountain ranges near the southern and western coasts of India. When it hits those areas, especially places like the Western Ghats in the southwest, it rains hard for the months of June and July. By early October, the monsoon rains are gone.

More than a billion people depend on the monsoon for their water. Scientists and government officials have said for decades that the monsoon is one of the most stable events in the natural world—never really varying more than 10 percent or so on either side of the equation.

In the occasional year where rainfall is more than 10 percent below what people are used to, it's a "bad" monsoon season. When rainfall is more than 10 percent above what people expect, then it's a "good" monsoon. This stability is critical to India, because it receives 80 percent of its rainfall during this season. Nearly half of India's population relies on agriculture, and therefore the monsoon, for their income.

"The monsoon will gradually spread across India by July 15, bringing cheer, hope, insects, relief from the heat, better farm output, GDP growth and lower inflation," says Shivam Vij, an independent journalist based in Delhi. "There is no aspect of Indian life, its politics or economy, which is not affected by the monsoons."[1]

Everyone in India knows what the monsoon season means to their communities. In a "bad" monsoon year—when there is a water deficit compared to historical averages—politicians can lose their jobs. People believe that economy-wrecking inflation is caused by "bad" rains.[2]

Conversely, a "good" monsoon can lift people's spirits, creating surges in consumer spending that can boost the economy. In a "good" monsoon, it "affect[s] things ranging from bank interest to the fortunes of the fertilizer industry," says Vij. "It will also alleviate the drinking water crisis in many parts by replenishing ground water."[3]

A crippling drought in a bad monsoon can devastate a regional agricultural economy. In 2015, in the midst of just such a drought in Marathwada, which was experiencing rainfall 40 percent below historical norms, more than a thousand Indian farmers committed suicide.[4]

Scientists have studied the complex intricacies of this vast global system of wind and water for decades, and we still don't fully understand all of its mechanisms. Theories range from the "jet stream theory" (involving upper atmosphere winds) to the "dynamic theory" (involving bands of weather that circle the globe). But what scientists and Indian government officials say, over and over, is that the long-term stability of the monsoon is never in question, no matter how many bad monsoons might show up occasionally.

This, from the Royal Meteorological Society, is typical. "The Indian monsoon is remarkably stable as a whole, with a mean total of around 850mm in the months of June to September, and an interannual [year-to-year] variation of only around 10% in most cases," it says in its overview of the monsoon.[5]

Indian meteorological and climate experts largely say the same thing. "If you compare India with other large countries in the world, the Indian monsoon is remarkably stable," says J. Srinivasan, who chairs the Divecha Centre for Climate Change at the Indian Institute of Science in Bengaluru. The monsoon can vary from year to year, Srinivasan says, but most of the time the total volume of rain is within 10 percent above or 10 percent below the long-term average.[6]

But cracks in the wall are starting to show up everywhere.

Scientists and statisticians have kept track of Indian monsoon rainfall levels since 1871, or for nearly 150 years. What this record shows is that major "drought" years (one standard deviation below the 10 percent deficit line) and major "flood" years (one standard deviation above the 10 percent surplus line) during that time were infrequent. There were just twenty-six major drought years in the past 150 years, and nineteen major flood years.

That's changed dramatically in the past fifteen to twenty years. One-third of the past fifteen years have been major drought years. There hasn't been a major flood year since 1994. Put another way, there have been five major drought years in India during the monsoon season in the twenty-first century. There were no major flood years in that time. This could simply be an anomaly inside a forty-year cycle—which has happened once before—or a trend that means something more ominous. India has essentially been in a drought cycle now for nearly half a century, the statistics show.[7]

Meanwhile, despite the apparent belief that nothing much has changed over the years, the total volume of rainfall from the monsoons, in fact, has been gradually falling over the past one hundred years.

Three Indian agricultural meteorology scientists from Kerala Agricultural University studied rainfall patterns over the state of Kerala (known as the "gateway of the summer monsoon") from 1871 to 2005. They found a "significant decrease in southwest monsoon rainfall" over time.[8] There was a slight increase in rainfall during the winter season, but it was more than offset by a much larger decrease during June and July, which are historically the rainiest months.

"The analysis revealed [a] significant decrease in southwest monsoon rainfall," the researchers wrote. "Rainfall during winter and summer seasons showed [an] insignificant increasing trend [while] rainfall during June and July showed [a] significant decreasing trend." Farmers planting crops might benefit from the slight increase in rainfall after the traditional monsoon season, they wrote, but the significant decrease over time in the two rainiest months (June and July) is a troubling trend.

Six scientists from the Sorbonne, the University of Maryland, and the Indian Institute of Tropical Meteorology found precisely the same pattern in a more recent study.[9] For their study, they analyzed multiple data sets that covered more than a century's worth of data (1901–2012). They, too, found a significant downward turn over that time period.

"There are large uncertainties looming over the status and fate of the South Asian summer monsoon," with scientists still vigorously debating whether the monsoon will weaken or strengthen over time this century, they wrote. "Our analysis using multiple observed datasets demonstrates a significant weakening trend in summer rainfall during 1901–2012 over the central-east and northern regions of India, along the Ganges-Brahmaputra-Meghna Basins and the Himalayan foothills, where agriculture is still largely rain-fed."

While earlier studies had largely predicted an increase in rainfall as land temperatures warmed, the data sets they observed showed the reverse. They concluded that a century's worth of data about actual rainfall patterns, combined with additional supercomputer modeling, showed that rainfall over South Asia was reduced by moderately rising land temperatures that collided with much greater warming in the Indian Ocean.

"Using observations and climate model experiments, we demonstrate that this reduction in rainfall is linked to the rapid warming of the Indian Ocean, especially its western part, during the past century," they wrote.

The effects are most clearly seen today in the Western Ghats mountain range in southwest India, where much of the water is deposited as it comes off the Indian Ocean at the start of the monsoon season.

"Research says rainfall is declining over the Western Ghats—and declining quite a bit. It differs in different places, but over the Western Ghats, as a whole, there is a decline," R. K. Chaturvedi, an expert in the Western Ghats at the India Institute of Science in Bangalore, told Al Jazeera. "That has been happening over the past 30 to 40 years. Kerala is really bearing the brunt of the situation, experiencing the largest of these declines. But it doesn't make headlines."[10]

The same group of scientists who found the substantial decline in monsoon rainfall over the past century also looked at *why* it had occurred. What they found was that, in fact, a literal black sky (combined with warmer ocean temperatures) was largely responsible for the substantial decline. Just as scientists were eventually able to show conclusively that smoking causes lung cancer through a complex physiological process, scientists are now finding that a black sky has unintentional, and potentially ominous, repercussions in South Asia. This, combined with rising temperatures and extreme rainfall events, may pose a triple threat to a monsoon season that hundreds of millions of people have relied on for centuries.[11]

"The World Health Organization has warned for more than a decade that rising air pollution levels pose a serious threat to human health worldwide, especially in developing countries, and high levels of pollution in the urban centers of China and India are now responsible for the premature deaths of more than 2 million people every year," said one of the lead authors of both studies, Raghu Murtugudde, an atmospheric science professor at the University of Maryland. "As if this news were not bad enough, my colleagues and I have found that pollution and dust particles blanketing that region are responsible for a 20-percent decline in South Asian monsoon rainfall over the past century."[12]

Scientists have believed for some time that rainfall would actually increase during the monsoon season as warmer ocean waters met warmer land, but this hasn't happened. The land temperatures haven't increased as expected, and monsoon rainfall has gone down substantially. Whether it will eventually increase, or whether the monsoon may be about to enter an unstable phase, is still an open question, but Murtugudde feels like he and his colleagues have solved at least part of the equation.

"Our new study . . . unravels this mystery by demonstrating that the increase in pollution has, in fact, slowed down the warming of the South Asian landmass," he said. "Meanwhile, the Indian Ocean has continued to warm rapidly, thus reducing the overall magnitude of the land-ocean temperature gradient. This weakening of the land-ocean contrast has weakened the monsoon."

Murtugudde and other colleagues also have discovered another disturbing trend in the "stable" monsoon over South Asia. Over the past fifty years, the monsoon has begun to start later and end earlier—depressing even further the total amount of rainfall and fresh water that is ultimately available, they found in two separate studies.[13,14] The monsoon season, on average, now begins closer to June 5 instead of June 1, and ends several days earlier than it once did a generation ago.

"This late onset and early withdrawal in recent decades has shortened the rainy season overall, robbing the region of even more life-giving rain," he said.[15]

Making the situation worse is the sobering news from research (based on NASA Earth-observational satellites) two years ago showing that the groundwater reserves in India are among the lowest in the world.[16] The Indus Basin aquifer, which straddles northwestern India and Pakistan, is the second-most stressed aquifer system in the world. Only the groundwater aquifer system in Saudi Arabia (which, by all accounts, has been largely depleted) is in worse shape.

"What happens when a highly stressed aquifer is located in a region with socioeconomic or political tensions that can't supplement declining water supplies fast enough?" asked Alexandra Richey, one of the study's lead authors. "We're trying to raise red flags now to pinpoint where active management today could protect future lives and livelihoods."

Researchers are still struggling today to understand how much water, in fact, remains in both of these huge aquifer systems. The truth is that no one knows how much water is left in the Indus Basin aquifer (or any of the other major aquifers). "Given how quickly we are consuming the world's groundwater reserves, we need a coordinated global effort to determine how much is left," said Jay Famiglietti, who is the senior water scientist at NASA's Jet Propulsion Laboratory.

In the midst of this new research confirming three serious long-term trends—a substantial downward trend in monsoon rainfall in the past one hundred years; changes in the start and end of the monsoon season; and a serious depletion of the major water aquifer system providing water to both India and Pakistan—scientists are also seeing an increasing pattern of "bursts" in extreme rainfall within an individual monsoon season itself.

In 2015, for instance, there was a deluge in June followed by a very small amount of rain in July and August, and then a bit more rain in September. Observers are noticing this sort of pattern more often. These concentrated bursts of extreme precipitation are difficult for the soil to fully absorb and for reservoirs to fully collect.[17]

Other scientists have confirmed this trend. One study, for instance, showed that extreme rain events during the monsoons—again, precisely what scientists say will occur more and more now—have increased over central India in the last fifty years, while moderate rain events have decreased.[18]

Confounding all of this is the relatively straightforward science, which has been established for some time now, that India has simply grown warmer over the past sixty years, making the downward and extreme rainfall trends even harder to adjust to for farmers and communities.

"Glaciers are melting in the Himalayas, and orchards in the range's valleys are being planted on ever-higher slopes in search of a temperate climate," *The Economist* noted. "Crops in the northern grain belt, notably wheat, are near their maximum tolerance to heat, and so are vulnerable to short-term blasts of higher temperatures. North India's cities are also growing hotter."[19]

While no one really likes to talk about it much, there is also a looming, existential question that hangs over all of this new research on declining monsoon rainfall, bursts of extreme rainfall, and changes in the start and end dates of the monsoon season. Could the monsoon "flip" abruptly—could it cross a threshold, or tipping point, and become unstable in a relatively short period of time?

This isn't the type of question any Indian politician ever wants to take on, but scientists have tentatively tried to answer it by looking at paleoclimate records. What they've found is that, yes, abrupt monsoon transitions over millennial (thousand-year) timescales have occurred in the past.

"Paleo-records from China demonstrate that the East Asian Summer Monsoon is dominated by abrupt and large magnitude monsoon shifts on millennial timescales, switching between periods of high and weak monsoon rains," scientists said in one such study, in 2015. These "abrupt" shifts, however, generally occurred over a period of hundreds of years in the geological record. The fact that there have been such big shifts in the past doesn't mean another one is imminent in the near future.

However, these researchers cautioned, there were clear signs right before monsoons shifted abruptly in the past. There were likely early signals that a monsoon transition was about to happen, which were only really evident afterward in the geological records.[20]

Still, some researchers have shown that very abrupt transitions in the monsoon season have occurred quite quickly. One study, for instance, showed that an abrupt shift in the Indian monsoon season happened just four thousand years ago. It shifted dramatically in only a decade or so and then remained that way for two centuries. The signatures of that shift—an earlier end to the monsoon season, as well as a downward trend in rainfall—are similar to what we're witnessing today.[21]

So, even though we may not know when—or if—the monsoon might be in serious trouble, there are clear warning signs right now that *something* is happening. For this reason alone, it makes sense to plan ahead.

Yet, despite the fact that major drought years have been happening much more frequently since the turn of the century, and despite the increased patterns of bursts of extreme rainfall that are difficult to deal with, water management strategies are still antiquated throughout India. No one is really prepared to handle the extremes in either direction (major drought years, or bursts of extra rainfall).

The good news is that, while we can't control the monsoon season, we *can* control how we respond if it is, in fact, changing substantially in the face of a new normal for Earth's system. Just as we saw in the Sahel with successful efforts by local farmers to hold back the Sahara's advances, local solutions in India will work here as well.

An Indian environmentalist, Anupam Mishra, preached the doctrine of local, traditional rainwater harvesting techniques—combined with a local community response—until his death in December 2016. Mishra's life work was documenting and arguing for the revival of traditional methods of community water management techniques across India.[22] Now, finally, it appears that people are willing to listen.

Mishra's method is this: for individuals to collect every last drop of scarce rainwater and for communities to build and operate their own local water reservoirs as a hedge against extreme drought years and extreme rainfall events—both of which are happening now and likely to increase even more over time.

Like the local farmers in Niger and Burkina Faso, who pioneered ways to capture scarce water at the base of trees embedded within crops, Mishra's method would appear to be precisely the sort of solution needed for India's coming water woes.

The types of local community solutions Mishra preached may be India's best hope in the near term. "Even the poorest farmers could work together better to store rainwater, for instance in ponds and tanks," *The Economist* wrote, "rather than praying for the skies to open."[23]

22

China

China approached Peru and Brazil with a proposition several years ago. It would build a three thousand–mile railroad from the western coast of Peru to the eastern coast of Brazil to handle commerce and trade from the interior of South America to China. It was an extraordinarily ambitious project.

If successful, it would give Brazil's soybean farmers a cheaper, more direct route to China than the cumbersome and increasingly expensive shipping route through the Panama Canal. The benefits for Peru are less obvious, other than the fact that China was willing to spend considerable sums of money on a massive infrastructure project that would certainly expand its trade options. China had become a hugely important trade partner for Brazil, and the railroad would cement that relationship.

And the benefits for China? Those are even less obvious and require a careful understanding of what the country faces over daunting food security and water scarcity issues in the next decade. China certainly doesn't talk about the connections between a railroad that snakes its way across South America and looming problems in northern China that could cause immense, near-term harm not only to its own population but to neighbors and the world's superpowers like the United States.

But the connections are there—in economic forecasts about the trade-offs that China is now forced to make over the types of food it grows inside its own borders and the types of food it must import from other places in the world . . . or in business sector reports about agricultural and food companies it's now buying in order to make certain that a countrywide famine doesn't disrupt the tight political and financial leadership that controls the country's position in the global community.

The China railroad across the northern half of South America immediately ran into environmental activist headwinds. Political leaders and business interests in all three countries wanted the railroad, but almost no one else did. Several routes were proposed for the ambitious project, but all would still cut a swath through the Amazon in order to reach Peru's coastline.

Environmental activist groups organized protests against the railroad that delayed plans for months on end. They argued, rightly, that the railroad would disrupt the delicate ecosystem balances in the Amazon's rainforest. The project would also likely accelerate deforestation in the Amazon by making it considerably easier for cattle farmers to ship to market as they clear-cut vast swaths of the Amazon.

Political chaos in Brazil delayed the project's plans for months as well. Authorities in both countries eventually signed off on plans, but it is still an open question whether the ambitious railroad through the Amazon will ever be built someday. China has shown considerable resolve in bringing such massive infrastructure projects to fruition. No one should ever doubt its ability to succeed where other countries might fail. But, for now, the Amazon railroad is still in the planning stages. Soybean producers in Brazil, Argentina, Bolivia, and Paraguay continue to ship their product via existing shipping and rail routes.

The much bigger question here is why China was willing to go to such extraordinary lengths in the first place. Yes, a three thousand–mile railroad through the heart of the Amazon would shorten the time that soybeans from Brazil might reach mainland China, and bypassing the Panama Canal to ship across South America and then from a Peruvian port would almost certainly save the Chinese money. But why the need in the first place? Are soybeans a genuinely strategic resource, requiring them to go to such extraordinary lengths to secure their continued supply?

The answer, in a word, is yes. Soybeans have become quite important to China. They are the answer—for now—to a looming crisis that has been building for twenty years and now threatens the fabric of the Chinese economy in the near future.

Virtually all of the world's soybeans—a crop used for food products but mostly as the basis of feed for livestock like pigs—come from the United States, Brazil, and Argentina. China used to grow its own soybeans—the soybean, in fact, originated in eastern China—but that has changed radically in just the past decade or so. Soybean meal is the largest source of protein feed in the world. For this reason, most of the world's soybean harvest is consumed indirectly by humans through products like chicken, pork, and beef.

In 1995, China imported just 18 million bushels of soybeans. It grew soybeans for use in food and livestock feed itself. Since that time, however, as China's economy has grown considerably and as its population has surged, China has become the world's largest importer of soybeans. Today, China imports more than 2.7 billion bushels of soybeans worldwide. That's quite a change—from 18 million bushels twenty years ago to 2.7 *billion* bushels today.[1]

Sixty percent of all soybeans grown in the world are now exported to China. Its appetite for the crop shows no signs of slowing down, with 5–8 percent growth per year.[2] Experts predict that this insatiable appetite could outstrip the entire global production of soybeans—everywhere, including in the United States and Brazil—within a decade. This partially explains why China is willing to build a railroad through the Amazon. It needs to buy almost every soybean grown throughout all of South America.

The American, Argentine, and Brazilian economies have all benefited from this strong, consistent demand in soybean exports to China. It's why trade disruptions (like a U.S.-China trade war threatened by President Trump) could cause serious harm to economies at both ends of the trading relationship. About half of U.S. soybean exports go to China, which is roughly one-third of the U.S. soybean crop.[3]

Why is China buying up the global soybean market? The steadily growing demand for soybeans in China is largely driven by its expansion of hog and poultry operations that use soymeal for feed. China's population is growing so rapidly that it has no choice; the soymeal indirectly feeds its population. Its middle class will double within a decade, and they are demanding meat products that match Western diets.

But China was also forced to make a serious trade-off in recent years that explains what's happening on the ground—and why it's buying up all of the world's soybeans now. This underlying threat, which China's political leaders don't talk about, or even truly acknowledge, is what is actually driving its relentless pursuit of soybean imports.

In northern China, where soybeans were once traditionally grown, water tables are now dropping at a rate of up to ten feet a year. Northern China (and parts of the west) is running out of water, and the water that's left is now so polluted in many of its rivers and streams that the Chinese government has an almost impossible pollution and sanitization threat on its hands.

As a direct result of this severe, ongoing drop in water tables in northern (and western) China, deserts are now expanding at a rate of 1,400 square miles a year. That is like adding a new desert larger than Rhode Island every single year. These drifting sands are covering cropland, making agriculture impossible.

This problem—desertification combined with dropping water tables—has made it almost impossible to grow soybeans in northern China. It takes 1,500 tons of water to produce just one ton of soybeans. China has ambitious plans to divert rivers from the south to irrigate the north, but those plans are in trouble now for ecological and political reasons. China is facing a massive agricultural and water challenge in the north right now. Soybeans are at the center of that story. By importing most of the world's soybeans, which use water from other parts of the planet, China is effectively importing a sixth of its water needs out of necessity.

China now finds itself squarely in the crosshairs of a wicked dilemma that is being driven by the very real, twin threats of Earth's changing climate and overuse that has polluted and drained its existing water supplies. Every leader in China understands the reality of this threat. It explains why the country is forced to make big infrastructure moves in remote places like the Amazon, even though it never explains these explicit threats publicly in any detail.

China's leadership began to recognize the triple threat of water scarcity, food insecurity, and a population explosion more than a decade ago. China had to make choices, and it chose to grow corn and wheat on shrinking, precious agricultural lands while importing soybeans it needed as livestock feed.

"China began looking overseas for external food sources in the mid-1990s, when it became clear that the country's production capacity for food, especially meat products, was insufficient to meet its rising demand," the Nature Conservancy wrote in a recent report on the Brazil-China soybean trade.[4]

Soybeans are currently the most important piece of agribusiness in Brazil, representing 10 percent of the country's total exports each year. Brazil's agribusiness is substantially dependent on China, just as China is dependent on Brazil's ability to grow and export soybeans to feed its own population.

"When evaluating which commodity to import, soybeans make more economic sense in a country with a significant shortage of agricultural land, since corn has a greater yield per hectare," the report said. "By 2009, soy represented 31% of the total Brazilian exports to China. From 2000 through 2009, Brazil's overall soybean exports rose from $2.2 billion to $11.4 billion. While this five-fold increase in total soybean export value is impressive, China's import of Brazilian soybeans by volume has increased nine times over between 2000 and 2010."

The choices that China's leaders face today are stark ones. China can't realistically grow soybeans in the north any longer as the desert swallows up the countryside. It desperately needs the soybeans as feed for livestock demanded by an exploding middle class. Any efforts to slow down the trade would likely lead to a middle-class revolt in China.

"While China was once the top exporter of soybeans, increased demand for soy combined with decreased production forced China to switch over to become a net importer in 1995 and by the year 2000 China had become the main global soy consumer and importer," the Nature Conservancy said in its report.

By 2009, China was importing 50 percent of the soy exported globally. Today, it's two-thirds. "There is no end in sight to this trend. According to the hedge fund Passport Capital, China would need to cultivate an additional area the size of Nebraska in order to meet its current soy demands," the report stated. "This scenario is unlikely considering that China's arable lands and water supply in the main soybean production zone are rapidly diminishing."

What China's leadership never talks about publicly is the way in which water scarcity and food insecurity are forcing its hand—and could potentially lead to resource wars with its neighbors in the near future. Desertification in the north and rapidly diminishing water supplies evidenced by water tables that are dropping year in and year out threaten China's environmental, economic, and even political stability.

For now, China is solving its food problems exacerbated by extreme water scarcity issues in the north by forging massive trade deals for soybeans and even livestock in places like the United States and Brazil.

Though largely unnoticed by the American public, a Chinese firm bought Smithfield Foods in Virginia—the world's largest pork producer—in 2013 as a hedge against its food insecurity threats. There was some concern locally—the firm employs nearly fifty thousand people in various parts of America—but China has gone out of its way to calm those fears.

"When a Chinese firm bought the world's largest pork producer, Smithfield Foods, two years ago, critics worried about U.S. employment, food safety and other issues," the U.S. government's public media arm, Voice of America, reported in 2015. "After two years under Chinese ownership, the Smithfield Foods company is doing well financially and employment is growing."[5]

But its purchase of Smithfield is a perfect illustration of the strategic moves China must make as its ability to feed its own population is diminished by the encroaching deserts in the north. Some experts fear that the Gobi Desert could swallow up the north and merge with other major desert regions. That, in turn, would put enormous pressure on China's leadership to seek other ways to obtain fresh water from the Himalayas, causing harm to Tibet and other countries.

The primary threat to thousands of farmers south of the Gobi Desert is the changing climate, Wang Shaokun, a researcher who studies desertification in Inner Mongolia, told an investigation journalism team in 2013. Declining rainfall, which causes the severe drop in water tables, is an immediate threat. "Our biggest concern today is not man-made problems, it is climate change and water resources," Wang said.[6]

As we've seen in Syria, Yemen, and elsewhere, when the farmers leave, they "may need to abandon their pastures and move into cities like Naiman or Tongliao. Leaving the land would be a sad fate for Mongolians . . . who have lived here since the invasion of Genghis Khan," the team reported on PRI.

China is monitoring its northern region, and the source of fresh water that originates in the Himalayas, closely. If China should choose at some point—perhaps in the next decade—to divert water resources from the Third Pole that embraces the Himalayas' system of water sources feeding major rivers, it will have immediate, profound implications not just for Tibet but India and Pakistan as well.

A *Seattle Times* team investigated the massive desertification situation facing China as well and found that the region was on the edge of collapse without significant infrastructure moves by China's leadership.[7]

"Many villages have been lost. Climate change and human activities have accelerated the desertification. China says government efforts to relocate residents, plant trees and limit herding have slowed or reversed desert growth in some areas. But the usefulness of those policies is debated by scientists, and deserts are expanding in critical regions," *The Times* reported in 2016.

"Nearly 20 percent of China is desert, and drought across northern China is getting worse. One recent estimate said China had 21,000 square miles more desert than what existed in 1975—about the size of Croatia," it reported. As one desert in the north expands, "it is merging with two other deserts to form a vast sea of sand that could become uninhabitable."

China has long recognized the threat from growing deserts. Decades ago, it tried to address the problem by pledging to plant tens of billions of trees in the north—a gargantuan reforestation project known as the "Green Wall of China." It is still ongoing and slated to run through 2050. But the project has been a colossal failure and has likely made the problem worse rather than better, experts say.

"While the intentions of this audacious project were noble, the lack of a proper environmental assessment, and an over-emphasis of planting quotas has actually exacerbated, not mitigated, the problem," said one expert risk analysis report.[8]

"The government introduced fast-growing, but non-native species such as pine and poplar, while simultaneously rooting out local keystone species like sea buckthorn during the 1980s. The removal of sea buckthorn, removed a species playing a vital role in holding the soil together, thus increasing erosion," the Global Risk Insights report stated. "The introduced pine and poplar are also very thirsty species, and introducing billions of them into an already arid environment, sunk the water table up to ten times below its original depth. This in turn killed off the shorter roots of prairie grasses, causing further desertification."

The abject failure of the Green Wall project has had an unfortunate and debilitating side effect as well—monster sandstorms that threaten air quality in some of China's major population centers. "The growing deserts have also added a new dimension to China's pollution problem, as giant sandstorms descend on the country from March to May, affecting cities such as Beijing, itself only a few hundred kilometers from the encroaching desert," according to the Global Risk Insights report. "Sandstorms not only add a new subcategory for hazardous air quality, but also pick up contaminants in the polluted soil of China's industrial north, raining toxic dust on major cities."

The end result is that desertification and water scarcity are worse than ever and loom as immediate threats. This is why the Dalai Lama once said privately to American officials that the greatest threat facing peace in the region was a direct result of Earth's changing climate and its immediate implications for China's security. Water scarcity and food insecurity were time bombs waiting to explode in the region, he said, according to a classified diplomatic cable released by WikiLeaks.[9] In fact, the Dalai Lama told the new U.S. ambassador at the start of President Obama's administration, he believed that any political discussions about the fate of Tibet should be put off for five to ten years while the more pressing problems caused by Earth's changing climate could be addressed.

"The Dalai Lama suggested the U.S. engage China on climate change in Tibet, recognizing that Tibetans could wait five to ten years for a political solution," the American embassy reported back to Washington in that classified diplomatic cable. "The Dalai Lama argued that the political agenda should be sidelined for five to ten years and the international community should shift its focus to climate change on the Tibetan plateau. Melting glaciers, deforestation, and increasingly polluted water from mining projects were problems that 'cannot wait.'"

What the exiled Tibetan spiritual leader feared, the embassy official reported, was that China would make moves in the region that threatened fresh water supplies for the entire region of countries that depended on the Himalayas for their own supplies.

Though China's leadership never talks about it publicly, entire cities in China are running out of water regularly today. The PBS show *Marketplace* found such a city recently—a small village in northern China called Lintao. They interviewed Yang Shufang, a resident in a wealthy, high-rise apartment, about the situation.[10]

"Yang doesn't live in the remote countryside, and her water isn't from a village well. She lives on the seventh floor of a luxury condominium complex in Lintao, a Chinese city with nearly 200,000 people that's run out of water," *Marketplace* reported.

Lintao is in northwest China, alongside the Tao River, which is a tributary of the Yellow River that originates in the Himalayan region. "The combination of a drought and a surge of urban development means the city's underground water supply has dwindled to dangerously low levels, leaving tens of thousands of people without easy access to the precious resource," *Marketplace* said.

Experts fear Lintao could be a sign of things to come. A previous Chinese premier, Wen Jiabao, once called the water scarcity problem in northern China "a threat to the survival of the Chinese nation."

"Four hundred Chinese cities now face a water shortage. One hundred and ten cities face a severe water shortage. This is a very serious problem," Liu Changming, a retired hydrologist for the Chinese Academy of Sciences in Beijing, told *Marketplace*.

If China's leadership makes the decision in the next decade that it must protect dozens—or hundreds—of cities in the north and northwest that are facing the loss of water, it may decide that it has no choice but to divert fresh water from the many tributaries that originate from the Himalayan region. If it does, it's impossible to predict how China's neighbors will react. India and Pakistan, at a minimum, will not stand idly by if China makes moves designed to divert water supplies.

"The biggest damage could be political," *The Economist* said in a special report on the water problems now confronting China.

Proposed dams on the upper reaches of the Brahmaputra, Mekong and other rivers are bound to have an impact on downstream countries, including India, Bangladesh and Vietnam. The Chinese say they would take only 1% of the run-off from the giant Brahmaputra.

But if all these projects were operational—and the engineering challenges of one or two of them are so daunting that even the Chinese might balk at them—they would affect the flow of rivers on which a billion people depend. Hence the worries for regional stability. And all this would increase China's water supplies by a mere 7%. The water crisis is driving China to desperate but ultimately unhelpful measures.[11]

For now, China is diverting water from the south in three large waterways projects. Two of them are completed, at a cost north of $60 billion. But even these projects will only marginally improve the north's water fortunes—and residents in the south aren't happy about the diversion as it stands. Experts are now discovering that water tables are dropping in the southern regions of China. Given this, there's not much else China's leadership can do to ease the north's water scarcity issues.[12]

Water is now China's Achilles' heel, and the situation has reached a breaking point. Water is China's "worst" problem, Wang Tao, from the Carnegie-Tsinghua Center in Beijing, told *The Economist*, "because of its scarcity, and because of its pollution."[13] Beijing has the same water scarcity issues that confront Saudi Arabia, Yemen, Syria, and others today.

"To fight for every drop of water or die: that is the challenge facing China," Wang Shucheng, China's former water minister, once said.[14] He also said the combination of desertification and current rates of water extraction meant that many cities in northern China—including Beijing, home to more than twenty million people—would run out of water in fifteen years.[15]

Shucheng made that prediction in 2005—thirteen years ago.

23

Environmental Diplomacy

For a very long time, the CIA, the NSA, and almost every branch of the intelligence community in the United States assessed emerging military threats and conflicts around the world based on the types of weapons systems a given country might possess, the percentage of its national budget it devoted to the military, and its political willingness to use those systems against opponents or internal rebel forces.

They gathered strategic intelligence from human intelligence sources inside countries (sometimes from spies, sometimes from diplomatic counterparts willing to trade information on a quid pro quo basis) and from reports, studies, or even newspaper articles that exposed strategic weaknesses or emerging military strategies.

The Pentagon's budget now is roughly $600 billion a year.[1] President Trump wants to increase it another $50 billion or so a year. It is the largest military budget in the history of the world and larger than the entire budgets of most countries. The U.S. spends as much on the military as Russia and China combined and is funding more military spending than at any time since the Second World War. The military budget is now bigger (in current dollars) than it was at the height of the Vietnam War.[2]

Putting aside what the U.S. military spends all of that money on around the world, the bigger question that almost never gets asked on Capitol Hill or anywhere else is this: How does the Pentagon know where to deploy its forces and might? And the related question: Does it always go in after things have gone south, or are there better ways to know where armed conflict might erupt?

As we've seen over and over, we now know—with a depth and level of certainty that we didn't have just twenty years ago—where conflicts over water scarcity, food insecurity, extreme flooding, massive superstorm disasters, multiyear droughts, and water or air pollution are imminent and almost certain to lead to conflict.

We saw it in Yemen a full two years before the water riots led to civil war. We saw it in Egypt, in Syria, in Pakistan. The only places where natural resource conflicts aren't turning into armed conflict are in countries such as Somalia, where everyone is hanging on for dear life and trying not to starve.

It's a critical question, one that needs to be asked, early and often. Yet it's rarely, if ever, asked. In North Korea, for instance, we're obsessively focused on its ability to put a miniaturized nuclear warhead on top of an intercontinental ballistic missile that can strike the U.S. mainland. We're willing to consider preemptive strikes against tests for both of these components of a long-range nuclear weapons system that North Korea does not (yet) possess.

But what does the Pentagon—or the American intelligence community— think about North Korea's massive food insecurity problem? Is it ever truly a consideration in its strategic considerations? The United States has traded food for peace with North Korea in the past, but it's always a bargaining chip, used at the last minute to avert the latest test of wills with North Korea's paranoid, conspiratorial dictatorship.

North Korea goes through famines on such a regular basis now that the outside world is largely immune to its pleas for help. We are willingly blind to the underlying reasons—and why this quite real problem for North Korea's leadership may, ultimately, hold the key to a more lasting peace in Asia.

Less than 20 percent of North Korea's land is now arable. In the past two decades, what's left of its ability to grow food on this shrinking environmental landscape has been whipsawed between increasing drought and floods, just as we've seen in other parts of the world.[3]

Out of desperation, North Korea's military leaders made foolish, and ultimately destructive, decisions to plant crops anywhere that there was open space. Farmers planted crops on steep hillsides, for instance, which destroyed forest cover and caused enormous erosion problems. The crop yields were small, while the environmental degradation was considerable.

When droughts lasted longer than usual, and extreme precipitation led to more flooding, the end result is what we might expect—famine at punishing levels for the country's twenty-five million people on a recurring basis. Even in times of plenty, where drought and floods aren't occurring, North Korea still requires substantial food aid.

On top of this dire situation, the nation's self-imposed isolation from the rest of the world means that it can't generate enough foreign exchange and capital to buy and import enough food to feed its people. As we've seen, relatively wealthy countries like Saudi Arabia and China have managed to pivot and use food imports to deal with the substantial changes in their environmental landscape.

North Korea has no ability whatsoever to pivot to food imports—not now or at any time in the foreseeable future as long as its government remains a dynastic dictatorship. But its problems aren't going away; they will only grow worse from here on out. Nearly half of the country's children under the age of five are malnourished. A large percentage of pregnant women are as well.

Public health advocates in other parts of the world focus a great deal of time, money, and effort on solving the associated severe medical problems that are a direct result of malnourishment. But big NGOs can't take the risk of working broadly inside North Korea on this kind of life-saving work, because virtually every aspect of their work is controlled by the country's military dictatorship. UN agencies, especially UNICEF, have a working relationship with North Korea that they enforce across the board: no access, no aid. As a result, huge swaths of North Korea have skyrocketing malnutrition rates because the country doesn't grant access for food aid.[4]

The most recent assessment from the United Nations is that two in five North Koreans now are malnourished. More than two-thirds of its people rely on food aid to survive. "Humanitarian needs [have] been exacerbated by 'recurrent natural hazards' such as frequent floods and drought," the BBC reported.[5]

The West, led by the United States, has imposed crippling sanctions on North Korea for years. The net result, say NGO advocates who have left the country and are free to talk about what they saw inside the country, is that the sanctions have deprived and depleted the UN agencies and NGOs that do try to operate inside the strict boundaries set up by the military. The sanctions haven't (yet) led to civil unrest that might force a regime change.[6]

Given all of this, it's been apparent for years that a genuine assessment of environmental threats in places like North Korea would make a great deal of sense for military planners and intelligence community analysts. It's entirely possible, for instance, that a concerted effort to address the underlying problems of North Korea's chronic food insecurity problems might have a lasting impact in a way that military threats and economic sanctions can never achieve. Such an effort requires a real, clear-eyed look at the geopolitical implications of the ways in which the changing environmental landscape impacts countries like North Korea.

President Obama's senior military and national security advisors reached precisely this conclusion early in his administration. The decision wasn't a partisan one. The Obama White House was following up on a national intelligence assessment of the national security implications of a changing climate system (and the environmental implications) published during President Bush's second term in office.

Obama's National Intelligence Council, which serves as the intelligence community's center for long-term strategic analysis, began to systematically study six regions of the world where environmental impacts like water scarcity, drought, and food challenges will have national security implications in the next ten to fifteen years.

The NIC is the bridge between the intelligence and policy communities (e.g., the White House's National Security Council).[7] Its strategic analysis reports are compiled by national intelligence officers, who are pulled in from government, academia, and the private sector. It's the group that provides the intelligence community with deep, expert analysis on issues that are at the top of the military and national security community list of problems around the world.

The six regions Obama's NIC set out to study in depth were India, China, Russia, North Africa, the Caribbean, and Southeast Asia. It followed up the assessments in these six regions with broad research reports on global food security, agricultural technologies, wildlife and conservation threats, and water scarcity issues.

For instance, the NIC's Southeast Asia analyses, coordinated under contract with the CIA's office of the chief scientist, focused on important flash points where moves made by governments in response to emerging environmental risks or natural resource constraints would almost certainly lead to conflict.

Dam building on the Mekong River Basin, for instance, "poses a potential catastrophic threat to agriculture, fisheries, and human habitation in Cambodia and Vietnam's Mekong Delta," the NIC's report said.[8] This is especially critical in Vietnam, where the Mekong Delta is responsible for half the country's population.

Deforestation and overfishing were likely to "bring about the near-term collapse of vital regional ecosystems, including the tropical forests and the fisheries of the South China Sea," it said. Massive wildfires (another climate signal) "threaten the environment and public health across the [entire] region."

Nearly all of the largest cities throughout Southeast Asia will face serious water management issues, including Bangkok, Dili, Kula Lumpur, Manila, and Singapore. Food insecurity is an existing threat in the Philippines, Laos, Cambodia, Burma, and Indonesia.

One of the NIC assessment's central conclusions is stark: "Disruptions to traditional lifestyles, water and food stress, and more frequent or more severe natural disasters will destabilize Southeast Asian societies . . . the poor, ethnic and religious minorities, and those living in peripheral areas of states will suffer disproportionately."

The NIC report also said that countries throughout Southeast Asia simply weren't prepared for inevitable mass migration—which, as we've seen, is a common thread in virtually every country or region facing these challenges.

"Large-scale migration from rural and coastal areas into cities will increase friction between diverse social groups already under stress," it said. "The country most in need of massive resettlement planning is Vietnam."

It also said that several countries were at risk of collapse within fifteen years: "Laos, Burma, and Cambodia are most at risk of partial or complete state failure."

Some of what the NIC was anticipating at the start of the Obama administration is beginning to play out in these countries. Warming oceans make tropical storms and typhoons more intense—making them even more deadly and displacing even more people. Since 2013, nearly fifteen million people have been displaced by typhoons and storms in the Philippines. Typhoon Haiyan killed more than seven thousand people.[9]

The wild card in all of their scenarios was China. In nearly every instance, the way in which China dealt with its own resource questions—and how generous it would be toward refugees fleeing to it—would likely have a profound impact on the other countries throughout Southeast Asia.[10]

China is just as critical to North Korea and East Asia, where the immediate and direct effect of military conflict or civil unrest can swiftly escalate into nuclear war.

The NIC report's authors said they believed that changes in the environmental landscape would almost certainly lead to refugees trying to escape their circumstances. "China may face refugee inflows from Southeast Asia [and] North Korea," it stated.

While a flood of refugees might not immediately stress China's capacity to assimilate them, they'd probably be separated from the Chinese population, creating the potential for civil unrest.

"China is likely to try to keep refugee and immigrant groups geographically contained," the report said. "China has pursued a similar policy with the significant refugee flows it has received from North Korea, not only environmental refugees but also political and economic ones. North Korea's capacity to cope with climatic pressures is very questionable given the ruinous state of its economy. As has been the case with past humanitarian disasters such as famines, a climate change-induced catastrophe in North Korea would by default spill over into China's Northeast."[11]

An assessment of South Asia by experts at the UN's migration agency[12] in December 2016 came to many of the same conclusions on the threat of civil society collapse from environmental factors and the accompanying refugee problem.

In a report on the "environmental degradation and migration nexus" in South Asia, the International Organization for Migration said that South Asian countries were facing an unprecedented threat precisely at the moment where a seventh of the world's population was migrating.[13]

"The world today is witnessing an era of unprecedented human mobility with more than one billion people on the move," said Sarat Dash, the agency's chief of mission in Bangladesh. "Forced migration due to poverty, conflict, climate change, and disasters can lead to deterioration in development outcomes." For South Asia, he said, "the range of sudden and slow onset events like changing rainfall, rising sea-levels, coastal erosion, floods, salinity intrusion and droughts put communities at greater risk impacting their economic, health, food and security conditions."

The IOM report included fairly blunt assessments of what their countries were facing from the environment ministers from Bangladesh, the Maldives, and Nepal.

It is "evident that environmental degradation [is] influencing the migration of vulnerable people," said Kamal Uddin Ahmed, Bangladesh's environment minister. "Bangladeshis . . . are familiar with the challenges imposed by floods, droughts and many other environmental threats. However, the growing irregularity and recurrence of climate extremes . . . is leaving no room for the vulnerable people to sustain within their own capacities but to migrate to a safer place such as urban areas."

The Maldives' environment minister, Abdullahi Majeed, was just as direct. "The impacts of climate change are of utmost concern to the Maldives. Environmental change remains one of [our] key drivers of population migration," he wrote. "Coastal erosion, depletion of ground water lens, and damages due to extreme weather events" are key factors driving migration.

Nepal's environment minister, Bishwa Nath Oli, said his country urgently needed a national policy on environmental risk and migration. "Adaptation measures which address the challenges of environmental risks, should include migration, as movement in search of livelihoods and shelter, or towards safer places both internally and across borders, is persistent."

These three countries, in particular, face a multitude of threats that drive migration, ranging from cyclones and storm surges, salinity intrusion, coastal erosion, and flooding to heavy rainfall, glacial lake outbursts, and droughts.

Six million Bangladeshis have already been forced to migrate due to environmental factors. Millions more will migrate in the near future, the agency said.

The cost of "adapting" on the Maldives is prohibitively expensive, leaving migration of the population as its only viable option. Half of the Maldives' population now believes that migration is its only option left.

Nearly everyone in Nepal believes that economic uncertainty and poverty is what drives migration. Even as they've experienced direct impacts (like water scarcity that forces them to hike down mountainsides to find water), the people of Nepal don't yet associate those impacts with the economic uncertainty these impacts lead to and create.

In virtually all of these countries, migration is essentially an adaptation of last resort in the face of water, food, or land problems that are just now starting to make their presence known. But it's in North Korea where these problems can trigger the kind of military conflict that no one wants.

The North Koreans who somehow manage to escape their dictatorial regime know what they're up against. In a survey conducted by the Korea

Environment Institute (South Korea's state-run institute that studies environmental policies), more than 70 percent of North Korean defectors said that they had experienced a significant climate event (like a prolonged heat wave or heavier-than-usual rains).[14]

Even North Korea's state-run media reported in 2015 that it had recently suffered "the worst drought in 100 years." And Kim Jong-un, in a rare public admission, seemed to acknowledge the threats as well—but blamed it on lousy equipment. During a field trip to its state-run meteorological center, Kim said there had been many "incorrect forecasts as the meteorological observation has not been put on a modern and scientific basis," according to the official Korean Central News Agency.

Given that this is one of the very few instances where the imperious North Korean dictator has been willing to display any sort of weakness whatsoever—state-run media often portray its leader as someone with near-superhuman capabilities, like learning how to drive a car at the age of three or notching a hat trick against an NHL all-star team as a goalie—it is logical to assume that North Korea's bleak environmental future may, in fact, create an opening that doesn't exist elsewhere.[15]

Foreign policy experts who have spent years watching North Korea's succession of dictators now believe that Kim Jong-un may be acting somewhat rationally in holding on to his nuclear arsenal. *The Economist*, in a 2017 cover story on this new thinking, makes precisely that point.

"For all his eccentricities, Mr. Kim is behaving rationally. He watched Muammar Qadaffi of Libya give up his nuclear program in return for better relations with the West—and end up dead. He sees his nuclear arsenal as a guarantee that his regime, and he, will survive. (Though it would be suicidal for him to use it.)," *The Economist* wrote. "Economic sanctions that harm his people will not spoil his lunch. Cyber-attacks, which may account for the failure of some recent missile launches, can slow but not stop him. America can solve the Korean conundrum only with China's help."[16]

If true, it means there may, in fact, now be a meaningful diplomatic opening on the environmental engagement front. North Korea's complete inability to feed its people now—the end result of an environmental landscape that is no longer capable of supporting them—has made that discussion a question of survival now for millions of North Koreans.

"Environmental engagement with North Korea might open up a window for engagement that has been slammed shut by fractious denuclearization politics," Benjamin Habib, an expert in the nexus between the environment and national security, wrote in *The Diplomat* in October 2016. "Environmental vulnerabilities matter, because they could threaten the stability of the Kim government. This vulnerability gives North Korea a stake in advancing environmental cooperation based on its own self-interest."[17]

Long before that can happen, however, U.S. leaders first need to recognize the genuine environmental risks that North Korea and other countries in Asia are facing and factor them into regional national security calculations.

Presidents Bush and Obama began to move in that direction a decade ago and directed their national security and intelligence community research teams to consider the emerging nexus between environmental risk or natural resource constraints and strategic national security interests. Whether that effort continues now—in a political atmosphere focused on closing borders and rapidly scaling military capabilities—is an open question.

Global tensions are set to grow as the resources necessary for life dwindle. Countries that can afford to expand businesses internationally, such as Saudi Arabia and China, pay to grow crops elsewhere and use those water resources, but developing countries—such as Yemen, Pakistan, and Somalia—are left with no choices and become fertile breeding grounds for rebels and terrorism. Countries like North Korea will continue to flex their muscles to ensure their survival.

At the root of all these conflicts is a lack of fresh water and arable land. Climate change is already accelerating the problem. The sooner we realize that globally, the faster we can establish an innovative, solution-oriented type of international diplomacy that can work in the real world where all are impacted by climate change.

PART 5

The Blueprint

We now find ourselves in a box canyon as a planet. But there is a realistic way out.

In the next ten to fifteen years, life as we know it will start to change irreversibly. With CO_2 levels rising dramatically, threatening human existence, climate change is no longer just an "issue." It's a critical reality that impacts every single inhabitant on this home called Earth. With that said, how can we effect lasting change with such an overwhelming crisis?

The answer is in the opportunity that the crisis of climate change has created—one unparalleled in the world's history. Over the next fifteen years, necessary infrastructure shifts will cost $90 trillion, most revolving around energy needs. We're going to spend that money regardless, but how and where we spend those dollars is what matters.

"The Blueprint" explores the current economic transformation and offers a realistic, doable, and innovative plan—one that works with nearly every sector of the economy, each country at some level, and relies on individual decisions rather than international treaties. It's a path forward that prevents the worst impacts of climate change, provides sustainable energy—enough for the world's energy requirements—and could simultaneously lead to the greatest economic resurgence in history.

24

A Path Forward

Like President Roosevelt, who at first saw Albert Einstein's warnings about Nazi scientists producing the first atomic bomb as a theoretical threat, some Americans today view climate change as a theoretical "future threat." They put it far down the list of everyday concerns. Others aren't quite sure how much of it is real. Yet thousands of scientists agree that these seven basic scientific facts are now irrefutable:

1. Levels of greenhouse gases in the atmosphere have risen to levels never seen in human history.
2. Temperatures are going up. The years 2014, 2015, 2016, and 2017 were the hottest in human history.
3. Ice sheets are melting, and sea levels are rising.
4. The patterns of rainfall and drought are changing. Springs are arriving earlier.
5. Heat waves are getting worse, along with extreme precipitation.
6. The oceans are acidifying, threatening vast parts of the food web ecosystem.
7. On every continent and in every ocean, animals and plants are moving toward the poles.[1]

Thousands of peer-reviewed science papers present overwhelming evidence that, for scientists usually skeptical by nature, climate change is a "settled fact." Human activity since the start of the Industrial Revolution is the cause. Though companies and nations are making moves to lower carbon dioxide emissions and fossil fuel consumption—reducing greenhouse gas levels—the efforts are disparate and disconnected. None, in isolation, will help solve our planetary emergency.[2]

As Charles Keeling discovered, carbon dioxide levels have risen from 310 parts per million in the 1950s to more than 400 parts per million half a century later. The rate of increase of CO_2 levels is now doubling every decade. There is no precedent for this in recorded human history. The last time levels were this high, it presented significant challenges to every species on Earth.[3]

We have no choice now. We have to work together on real-world solutions. After decades of failure to act at a global level, civic, business, advocacy, consumer, and important world leaders are finally taking steps that scientists have been demanding since the 1980s. A blueprint is emerging—one that shows what we need to do to literally save the planet and its species, including human beings. We just need to learn how to follow the directions over the next ten to fifteen years to present this slow-moving catastrophe from altering life as we know it.

We need to leave four-fifths of our known fossil fuel reserves in the ground and significantly change the way we build out $90 trillion in new infrastructure in the next fifteen years to make certain we don't cross a potential tipping point of 450 parts per million of CO_2 in the atmosphere—a critical threshold beyond which the climate system could grow increasingly unstable. Half the species on our planet could disappear. Dust bowls could become commonplace. Drought and monsoons could become the "new normal" in parts of the world. And, unfortunately, it would take the human species one thousand years—or longer—to reverse the damage. Lowering fossil fuel consumption and reducing greenhouse gas levels can't harm our planet, but allowing CO_2 and other greenhouse gases to reach critical levels in our atmosphere quite possibly could mean the end of life as we know it on our planet at some point.[4,5]

What's the first step in this emerging blueprint? It's an easy one, but also one that has become increasingly difficult in the public sphere in America. We need to change the nature of the conversation in important and meaningful ways around our energy sources and their impact on Earth's climate. They aren't two separate conversations, which has been the case for at least two decades now. Republicans are comfortable talking about energy independence and economic freedom. Democrats are quite at ease with the environmental and climate change discourse. But the truth is now obvious. These two conversations, which have run in parallel, are actually the same conversation. It's well past the time to merge these two into one conversation. Our lives depend on it.

What are the elements of that conversation? For starters, we need to clearly understand and identify the very real dangers to the planet and its species (including humans). We need to stop ignoring or confusing specific, current, unimpeachable scientific research of our changing climate—globally, nationally, and locally. We need to recognize the thousands of compelling, real-life stories that personalize the climate and energy story so that it's no longer an abstract, distant threat.[6]

In short, we need a conversation that changes the tired, worn-out, redundant climate conversation to one that is relevant to every person on Earth and bypasses the often political overtones of the subject; presents clean-energy solutions that are not only possible but already in the works; encourages the development of a master blueprint that all organizations, businesses, and individuals can use to lower carbon dioxide emissions (the U.S. and China, combined, create more than 42 percent of carbon dioxide emissions for the planet); and moves climate change from a debate (with fossil fuel providers on one side and environmentalists on the other) to a robust search for creative solutions that work.[7]

We also need to change the conversation from merely an "issue" or a potential future threat to a critical reality that impacts every single person on the planet. We need to turn away from "green" thinking (save the planet) to "blue" thinking (save humanity) and create a blueprint where we can all work together to save our future.

While the world may just be waking up to the climate wolf at the door, the good news is that the world economy has been transformed over the last twenty-five years and is now nearly perfectly suited to a rapid transformation of a massive $90 trillion shift in its infrastructure that has energy needs at its center.[8]

Computing, communications, biotechnology, materials science, and other fields are in the midst of technological revolutions, greatly expanding humanity's productive capacity. World output has more than doubled since 1990, accompanied by rising international flows of knowledge, trade, and capital, as well as by enormous structural changes. Developing economies have grown in importance, with their share of global GDP rising from just over a quarter to more than two-fifths over this period. The number of people living in urban areas has surged by two-thirds, to more than half the world's population.[9]

Developing countries—the poorest and most populous region of the world—have been at the heart of many of these changes. Middle-income countries' output has more than tripled since 1990, and low-income countries' has more than doubled. Growth accelerated not only in large emerging economies such as China and India but also in many smaller, poorer countries in Asia, Africa, and Latin America. In developing countries, the number of poor fell by nearly five hundred million just in the last decade— the fastest pace of poverty reduction for which we have data. But 2.4 billion still live on less than two U.S. dollars a day. There is now an opportunity to build on this experience to make further major gains in human well-being in the next ten to twenty years. But there are major risks that overshadow what is otherwise a very bright prospect in this emerging blueprint for action at every level of this new global economy.[10]

In the wake of the Great Recession of 2008/09, countries are struggling to restore or achieve fast, equitable growth in output, jobs, and opportunities. Despite the rapid growth before the crisis, the world is not on track to eradicate extreme poverty by 2030. Improvement in broader measures of human development has also slowed since the crisis. Major recent natural disasters have inflicted significant economic and human costs, including Typhoon Haiyan in the Philippines, Hurricanes Sandy, Harvey, and Irma in the United States, major droughts in China, Brazil, and the Horn of Africa, and floods in Europe. Such extreme events are likely to increase in both frequency and magnitude with unchecked climate change. Nor are extreme events the only concern. Existing climate variability is already a major source of poverty and insecurity among the rural poor. For them, even small increments to risk in the form of delayed rain, higher temperatures, or slightly more intense or protracted drought can mean disaster.

Tackling the challenge of strong, equitable, and sustainable growth will require huge new investments and shifts in resource use. Actions today and in the next fifteen years will be critical to stabilizing and then reducing emissions to try to meet the targets the scientific community has established to keep us below a critical tipping point in the atmosphere. They will either lock in a future with inefficient infrastructure and systems, growing pollution, and worsening climate change or help move the world onto a more sustainable, low-carbon development path that strengthens resilience and begins to slow and reverse the accumulation of climate risk. That blueprint is now clear.

There are three fundamental drivers of this new, emerging road map: more efficient resource use, infrastructure investment, and innovation. Socioeconomic systems that hold the key to multiple economic, social, and environmental benefits (cities, land use, and energy systems) are all currently undergoing massive shifts. These systems are crucial in the next ten to twenty years, because they are so important for the global economy and emissions, and they are already undergoing rapid change. As a result, nearly all of the actions that are needed to keep us safe are, in fact, compatible with goals of boosting national development, equitable growth, and broadly shared improvements in living standards. The emerging blueprint looks like this: We can grow economies, lift hundreds of millions of people out of poverty, and distribute equitable economic growth to nearly every corner of the planet.

It is a moment like no other in human history, with a crisis creating an unparalleled economic wealth opportunity.

Reforms will entail costs and trade-offs and will often require governments to deal with difficult problems of political economy, distribution, and governance, but an argument that tackling climate risk is simply too costly—whether in terms of growth, competitiveness, jobs, or impact on the poor—has been substantially overstated (for political reasons), especially when the multiple, extensive, and mutual benefits of climate action are fully taken into account.

While there is not a simple formula or one-size-fits-all agenda that will work for all countries, the emerging blueprint outlines actions that are perfectly compatible with nearly every sector of the economy and each country at some level. It does not rely on a complicated international treaty. It doesn't rely on an act of Congress. Instead, it relies on individual decisions in every sector at the core of a $90 trillion shift in infrastructure—one that is occurring regardless—to follow a blueprint. Each will deal with development and climate challenges differently, based on levels of economic, human, and institutional development, social and political structures, history, geography, and natural endowments. Countries will need creative experiments in order to find the right path for their own circumstances.

We are going to invest $90 trillion in necessary infrastructure shifts over the next fifteen years no matter what. Most of it will revolve around energy needs. A tectonic shift in the world economy has created the opportunity. The emerging blueprint makes the opportunity before us crystal clear. How we act on that blueprint is what matters now.

25

Disruption

We are in the midst of an economic transformation unparalleled in the world's history. It's happening so quickly, and at such scale, that it's almost unrecognizable to most people.

That sea change is also enmeshed in deep, polarizing, conspiratorial political fights, further obscuring its rise. A number of the wealthiest corporations in the history of the world—and some government leaders at their beck and call—are bitterly opposed to this transformation and determined to deny it.

Yet it exists nevertheless.

Great economic transformations occur in every generation—some through speculation (the Great Depression), some through greed (the subprime housing market collapse), and some, as now, through necessity.

This new transformation will be seen most clearly in two sectors that directly impact hundreds of millions of consumers: transportation and utilities. By 2030, that transformation will be well under way.

Within just a decade or so, electric vehicles will begin to replace the internal combustion engine as the primary form of transportation in the United States and elsewhere. China, the largest new consumer market in the world, has said that only electric or hybrid vehicles will be allowed on the roads there by 2040.[1] At the same time, solar photovoltaics distributed across many places locally will have disrupted the utility sector providing our electricity and power.

The reason is simple. Driven by the necessity to find non-carbon sources of energy before we reach planetary tipping points created by industrial CO_2 that remains in the atmosphere for hundreds of years, renewable energy is in the midst of a revolution that is transforming the way the world produces and uses electricity.

The age of fossil fuels is actually coming to an end.

Solar, wind, geothermal, and hydropower are providing vast new energy resources, replacing carbon-intensive fossil fuels, and eliminating the greenhouse gas emissions that cause climate change.

The United Nations' Intergovernmental Panel on Climate Change (IPCC) has determined that, in order to avert the worst impacts of climate change, the world will need to limit global warming to 2 degrees Celsius below preindustrial times. The world has already warmed up 1 degree Celsius.[2]

We now know, with a high degree of certainty, that we can achieve this target by reducing emissions in every country and by rapidly scaling up the deployment of clean energy to completely replace the burning of fossil fuels by 2030.

Though this time frame may seem fast, clean energy is the latest in a long history of disruptive innovations—like television, the cell phone, and the personal computer—that grow exponentially, shift consumption paradigms, and create a new technological reality in a relatively short period of time.

It's happening now, almost everywhere.

The IPCC has calculated a global carbon budget that sets a limit to the remaining greenhouse gas emissions we can release for it to remain likely that we stay within the 2-degree boundary. In the simplest terms, the world is sending about forty billion tons of carbon into the atmosphere each year. We only have about five more years before we cross a threshold in that carbon budget that likely means the world will warm by 1.5 degrees Celsius.

We only have about twenty years left in our worldwide carbon budget before we hit the 2 degree Celsius red line.[3]

There is only one hopeful path forward for the world—only one, real blueprint that gives us any semblance of a chance to stay below 2 degrees Celsius. We need to slow carbon emissions as much as we possibly can in the next decade. But, at the same time, disruption and transformation in the energy and utility sectors need to occur at an unprecedented pace.

Accelerating the deployment of clean energy needs to supplant fossil fuels entirely, making it unnecessary to burn through the remaining carbon budget. The objective is to make substantial progress now and extend the budget "finish line" so far into the future that the target limit ultimately becomes moot. Not only does this buy time to forestall or completely prevent the worst impacts of climate change, it establishes a sustainable energy foundation that will produce more than enough cheap, reliable electricity for the world's energy requirements.

The core pieces of that renewable energy future are already around us. Today, the renewable power available in readily accessible locations from the three most common sources—sunlight, wind, and water—is more than fifty-five times the power needed worldwide by 2030, according to the U.S. Energy Information Administration (EIA).[4]

A 2009 study by Stanford University published in *Scientific American* used EIA data showing that the maximum power consumed worldwide at any given moment is about 12.5 trillion watts (terawatts, or TW). The EIA projects that in 2030 the world will require 16.9 TW of power by the current mix of today's fossil fuel-dependent sources. Global power demand, however, would be only 11.5 TW if the planet were powered entirely by renewables because of their higher energy efficiency than the burning of coal, oil, and gas, which waste much of their potential energy in the form of heat. Renewable power available in readily accessible locations from water (2 TW), wind (40–85 TW), and solar (580 TW) totals 667 TW, literally dozens of times more potential energy than the world will need by 2030.[5]

The renewable installations required worldwide would be a mix of photovoltaic (PV) power plants, concentrated solar plants, rooftop PV systems, wind turbines, and hydroelectric plants. The growth potential is huge since all of these technologies exist now at low cost, but currently have less than 1 percent of installations in place (except for hydroelectric dams at 70 percent).[6]

Solar power is at the epicenter of the clean energy revolution. The cost of manufacturing solar panels has been steadily falling for years, while the growth rate of utility-scale solar installations has skyrocketed. In 2014 alone, more than a third of all new electric capacity in the U.S. came from solar, which itself represented 41 percent growth over the year before. Solar energy is right now at an inflection point of technological proliferation and low cost.[7]

One of the fastest-growing areas is photovoltaic installation for residential, nonresidential, and utilities. The cost for rooftop solar panels has been dropping so fast that they are now common equipment for homes in many parts of the country. The global International Energy Agency (IEA) predicts that by 2050, PV solar alone will provide around 16 percent of total global electricity production, a significant increase from its 2010 prediction of 11 percent.[8]

Meanwhile, it is only a matter of time before electric vehicles surpass internal combustion engine vehicles. The electric car giant Tesla is leading the automobile transition, with plans to open a massive new factory in Nevada by 2020 that will manufacture a new generation of high-efficiency lithium-ion batteries that will power an affordable midsize sedan.

The so-called gigafactory will drive down the cost of its battery packs 30 percent by 2017 and 50 percent by 2020. With the gigafactory's 35 GWh of battery storage, Tesla plans to ramp electric vehicle production up to five hundred thousand cars by 2020. For perspective, Tesla says it produced thirty-five cars in 2014, with forty thousand Tesla cars already on the road today.[9]

Morgan Stanley predicts there will be 3.9 million Tesla cars on the road by 2028, which will be able to store 237 gigawatts of capacity. That's 22 percent of U.S. electric generating capacity and more than ten times the amount of grid storage capacity today.[10]

Perhaps more importantly, the new generation of high-efficiency, low-cost batteries will help wind and solar energy industries with their current problem of intermittency, or the reality that we need energy when the wind is not blowing or the sun is not shining. Better energy storage will help grow utility-scale renewable energy, along with the overall growth and diversification of the industry as a whole.

By 2030, then, we could see a radically different world on an accelerated timetable. But we are already well on our way to living in that world. China and the United States—the world's top two emitters—are also leaders in installed renewable capacity and technological innovation. In the first ten months of 2013, for instance, China doubled the pace of adding renewable energy capacity. China installed thirty-six thousand megawatts of hydro, solar, wind, and nuclear during that period and is on course to add more generating capacity from renewables by 2035 than the United States, Europe, and Japan combined.[11]

In 2013, more than a fifth of the world's electricity came from renewable sources, and clean energy accounted for over half of all net additions to global power capacity. This figure, published in REN21's *Renewables Global Status Report*, jumped by more than 8 percent overall in 2013, and total global installed capacity reached a record 1.5 million megawatts.[12]

Institutional investors now see this 2030 future clearly. JPMorgan Chase has committed $200 billion in renewable investments between now and then. Competitors like Morgan Stanley aren't far behind with similar commitments to investments in renewable energy, non–fossil fuel transportation, and energy efficiency that will disrupt markets in a matter of years, not decades. Current renewable investment is between a quarter and a half trillion dollars per year and is now set to skyrocket in the next fifteen years.[13]

Large utilities are also investing in renewables and distributed-grid technology on a large scale. Distributed-grid technology—like rooftop solar and small-scale wind turbines—has become so cheap and readily available that it is beginning to threaten the outdated business model of traditional power utilities. As more and more customers produce a large portion of their own electricity and invest in energy efficiency, utilities stand to lose $48 billion per year by 2025 while distributed-grid customers will retain most of that money.[14]

The goal of such a blueprint—essentially zero net carbon contributions worldwide by the year 2030—presumes major disruption of the current utility model for electric power and internal combustion engine for transportation. We are in the middle of the disruption. The International Energy Agency predicts that clean energy will receive almost 60 percent of the $5 trillion expected to be invested in new power plants over the next decade.[15]

Despite the fact that oil prices have fallen, demand for renewables like wind and solar power has continued to rise. Wind and solar have been growing by an average rate of more than 25 percent year over year. Bloomberg New Energy Finance predicts annual investment in new renewable power capacity will rise by between 250 percent and 450 percent by 2030.[16]

History has repeatedly shown that new technologies can upend established markets and rapidly transform the way people move around and communicate. The key to scaling up these breakthrough technologies is creating the infrastructure to sustain them.

The wireless telephone is a cornerstone example. The first wireless call was made more than forty years ago, and the first commercially available handset took ten years to reach the market, but the rapid adoption of mobile phones happened once the market for building the required infrastructure yielded the normal return on investment.

Wireless phone subscription was essentially zero thirty years ago. Today, smartphones are so ubiquitous that they are transforming the lives of people everywhere, and the biggest companies ship a combined billion units each year. Growth projections show that the majority of humanity will soon have a mobile phone, turning what was once an impossible idea into an inevitable fact in just a few years.

The history of the personal computer is another story of exponential growth that completely transformed markets. In the 1960s, computers were as big as an entire room, and only large institutions like universities and the military owned them. Personal computing was highly specialized, and most computers were appealing only for their novelty. Today, we engage in almost every function of modern life—shopping, traveling, working, communicating—using computers.

And some of the largest companies in the world like Apple, Google, Facebook, and Microsoft didn't exist a few decades ago. Today, their products and services are now integral to daily life. Indeed, innovative technologies like the personal computer insinuate themselves in everyday life because they transform existing markets by making consumption and communication easier, cheaper, or both.

Renewables are making one of our most basic activities—our use of energy—easier and cheaper. Distributed-grid technologies offer homeowners and small businesses the chance to generate their own power with zero fuel costs, and vast new economies of scale that drive down prices are making clean energy cheap, ubiquitous, and available to all.

The message is loud and clear: The future is here, now. Disruption is occurring in our biggest energy markets and within industries such as transportation that rely on those energy sources. The only question is whether that disruption will happen in time to stave off the worst impacts we're beginning to see all across the planet here and now as well—impacts, as we've seen throughout this book, that could end the world as we know it.

26

Waking the Behemoths

The human immunodeficiency virus (HIV) is not a partisan issue, though it once nearly became one until medical and pharmaceutical scientists untangled its deadly path. We don't question the scientists who discovered its true nature and, ultimately, found a way to contain it.[1]

Astronomy is no longer a religious issue, though it once was until it helped create the modern scientific era. We don't question the physical nature of our universe and Earth's place within it. When NASA lands a rover on Mars, we don't question the physics behind it.[2]

Nuclear power isn't magic, though the concept of splitting the atom and harnessing the energy released in such a controlled reaction may have seemed a bit like magic until the Manhattan Project physicists altered our understanding of the awesome power that can be unleashed with such knowledge.[3]

Climate scientists, who now universally understand the physical dynamics of how the modern industrial era is changing Earth's climate in dangerous and risky ways, are not somehow different from the other scientists we trust. They spent years acquiring their scientific expertise and knowledge, just like their colleagues in other science fields like medicine, astronomy, and physics.

Climate change needs to stop being a political issue. The fact that fundamental climate science is important to advocates in the environmental movement doesn't make it a political issue; it simply means that those people have paid closer attention to what the science says and means than others in the rest of society.

And just because former president Barack Obama, a Democrat, took climate change seriously and President Donald Trump, a Republican, manifestly ignores it doesn't make the issue inherently political. It simply means that Obama acted within a highly charged political system on the best available science, and Trump has chosen to largely ignore that same science.

The world's largest corporations—like Nestlé, for instance, which has chronicled and anticipated looming water shortages in nearly every corner of the earth—are now looking well beyond the partisanship that has marked this issue for nearly twenty years. Very large business forces capable of altering the fundamentals of the political system—and not necessarily represented by the U.S. Chamber of Commerce or the National Association of Manufacturers in Washington, D.C.—have finally entered the climate arena in new and meaningful ways.

While some political leaders are still trapped in a time warp—much like Catholic Church leaders were in Galileo's time on the question of whether the sun revolved around Earth, or the other way around—business leaders who don't represent oil or coal interests have stopped pretending that climate science is confused or uncertain.

Just as the scientific community has grown quite certain of the risks inherent in altering Earth's climate system, the global business community has now accepted the science of climate change and is acting on it in ways that might surprise people. Only the political system lags behind when it comes to the issue.

Nearly every large multinational corporation (even big oil companies such as ExxonMobil, Shell, Chevron, and BP) now accepts climate change science on its face. The only holdouts are coal companies that stand to lose most in a global energy transformation in the next twenty years, and private companies like Koch Industries that rely on the fossil fuel economy.

There are countless examples. The world's very largest corporations now recognize the urgency of what climate science and data is telling us and are committing resources to efforts to mitigate or adapt.

Let's start with Pepsi and Coca-Cola. They may creatively fight each other for world dominance in the "soda wars," but both are taking steps in various ways to respond to climate impacts affecting their supply chain business and bottom line.

The Coca-Cola Company, for instance, has created a comprehensive "field-to-market" environmental program using climate-related data to quantify water use, fertilizer use, energy use, and greenhouse emissions.[4] Half of Coca-Cola's global corn acquisition it needs to make its central consumer product is now part of this environmental program built around the reality of climate data.

PepsiCo announced the installation of a solar photovoltaic system that will supply massive amounts of renewable energy for the company's Gatorade manufacturing operations in Tolleson, Arizona. Pepsi officials publicly describe the effort as a way of preventing the release of fifty thousand tons of carbon and other greenhouse gases to the atmosphere. Pepsi has said it will use data from this solar project to help inform future solar installations and projects so it can reduce its greenhouse gas emissions globally.[5]

Two of the world's largest packaged food companies, Kellogg and General Mills, now have comprehensive environmental and climate data programs that span the planet.

General Mills' new, sweeping climate program deals with both mitigation and adaptation, across the board. It announced that it was setting global targets related to reductions in greenhouse gas emissions, energy, water, transportation, packaging, and solid waste.

"Business, together with governments, NGOs, and individuals, needs to act to reduce the human impact on climate change," the company said of its new climate policy. "As a global food company, General Mills recognizes the risks that climate change presents to humanity, our environment and our livelihoods. Changes in climate not only affect global food security but also impact General Mills' raw material supply which, in turn, affects our ability to deliver quality, finished product to our consumers and ultimately, value to our shareholders."[6]

The Kellogg Company has announced a commitment to use the Global Landscapes Initiative at the University of Minnesota—which openly shares data and maps that illustrate how climate change affects risks to major crops within the food system—in ways that will impact its global sourcing. Kellogg plans to use climate data to guide actions that help create efficient, adaptable, and sustainable supply chains.[7]

Three of the planet's largest nutrition companies—Mars, Nestlé, and Monsanto—have also announced comprehensive programs recognizing climate science impacts and ways to both mitigate and adapt as well as respond to global supply chain sourcing.

Mars Inc. has announced major scientific and climate data investments in food safety and plant science designed to create resilience across its agricultural supply chains and boost resource management and yields. The company also continues to invest in renewable energy, including plans for an enormous wind farm in Lamesa, Texas, that will offset the energy needs of its entire North American office and manufacturing footprint.[8]

Nestlé has established greenhouse-gas reduction targets based on science and has incorporated that data into its sourcing and supply chain decisions. Its "Farmer Connect" initiative is built around sustainability and water stewardship practices. Nestlé scientists are encouraged to use climate-related data for nutritional assessments in peer-reviewed scientific journals.[9]

Monsanto has announced a maize-breeding trial at multiple sites (using climate data from two of the world's largest research centers) to drive our understanding of how climate and water-availability changes will impact crop productivity and food security.[10]

The world's largest technology companies—Microsoft, Amazon, and IBM—are acting as well based on the latest science and climate impacts data.

Microsoft is sponsoring a series of efforts to use data tools to improve models studying climate preparedness, including a new Microsoft Research effort that specifically focuses on climate-related food resilience.[11]

IBM has announced a global grid program built to expand networked supercomputing efforts to study various climate change topics. Scientists studying topics such as water management or staple food crops will have access to up to one hundred thousand years of computing time (worth $60 million in today's dollars) under the program.[12]

Amazon announced a climate research grant program with a focus on computational analysis. The company's grants will cumulatively offer up to fifty million core hours through its own supercomputing resources (which are, as we can imagine, considerable). The Amazon offer is designed to accelerate our understanding of the scope and effects of climate change—and suggest potentially new ways to deal with it.[13]

And Walmart, the planet's largest retailer, has announced a plan to reduce its greenhouse gas emissions and substantially procure large amounts of renewable energy globally. Unlike the political system, which has steadfastly avoided setting any meaningful climate goals by 2020 and beyond, Walmart now has an actual 2020 goal of driving the production or procurement of seven billion kilowatt hours of renewable energy globally. It has also set firm goals to reduce the energy intensity required to power its stores globally by 20 percent compared to 2010 levels and has set up a data-driven index to measure, track, and identify key impacts in its supply chain, including greenhouse gas emissions.[14]

We are now at an inflection point—a pivot. The essentials of climate science are largely beyond dispute. The world's very largest businesses likewise are engaged and actively responding to climate science and data. They are reducing their impact and offering resources to further our understanding of climate impacts. If they view climate science, climate data, and climate impacts seriously, it's inevitable that the political system will also do so at some point.

27

The Anvil

America was forged on the anvil of two fiery processes—scientific peer review and the free market. When both inevitably lead to a single opportunity, American leaders have never failed to grab the hammer and create something that can remake the planet.

Until now.

Democratic leaders like, and implicitly trust, the scientific peer-review process. When something of value emerges from that fire—and, make no mistake, all can witness the flames from afar—they believe and step up to the anvil.

Republican leaders innately, deeply believe that the free market is the best and proper place to severely test raw elements—often in full, blinding, public display—of potentially game-changing, market-leading products that can drive innovation and economic development. When something is forged in this fiery process, they, too, take their place at the anvil.

For Democrats, the scientific peer-review process isn't merely the gold standard—it's the actual process that produces the gold. So, naturally, when gold emerges, they rush to shape it into a thing of beauty and value for the world to see. However, because they trust science, they sometimes overreach and use it in the service of political or environmental goals.

For Republicans, the free market isn't simply game-changing—it's the actual game itself. A government-driven command and control regime must be removed for products of immense value to emerge. It's how wealth is created, how new industries emerge, and how America leads.

When these two powerful forces coexist in America, greatness emerges . . . and nearly everyone benefits. It is a time-tested, historic fact. The evidence is all around us.

The internet was born this way. Google's initial million dollars came from the National Science Foundation (NSF) as part of a grant to Stanford for the first set of scientific experiments with digital libraries, but the market and the nearly unrestricted forces of capitalism allowed it to succeed and change the world. The first interconnected supercomputers were designed for science at the Defense Advanced Research Projects Agency (DARPA) and NSF but were handed off in a frenetic, unregulated rush to allow information to be shared freely across the planet. That fiery technology forge created the largest companies in the history of the planet, forever changing the global economy.

Today, the processes of scientific peer review and the free market have produced something of unmatched beauty on the anvil of democracy. That product is *climate change*, and it has the untapped potential to remake a depressed global economy (built around new energy sources) that has always responded to price signals.

The truth is that the climate science peer-review process is the easier of the two to understand. Every corporate CEO—even those at fossil fuel companies like ExxonMobil or Shell—now acknowledge its implicit truth. The fiery gold-standard peer-review process has forged a consensus in the complicated, messy science of climate change. A nearly endless series of pieces of peer-reviewed scientific evidence over the years makes it clear with no possibility of uncertainty that climate change is now threatening people and communities everywhere. The voices of uncertainty that still remain on climate change aren't scientific ones.

I've spoken to hundreds of business leaders in the past decade. What's stark—and not very well known to the public—is that the business community knows full well that the scientific peer-review process has quietly answered the clarion call on climate change. They know that climate change is real, that we are largely responsible for it, that it's changing our planet in dangerous ways, and that the uncertainties that may have existed twenty years ago no longer exist. It's here now, and boardrooms have already begun to build it into their assumptions and business plans.

This science- and evidence-based knowledge among American business leaders who implicitly trust the free market leads to the harder, and more complicated, question: What do we do about the fact of climate change that now affects nearly every part of what we see outside our windows at work and home . . . without damaging a fragile economy? We know that renewable energy needs to replace traditional energy sources as quickly as possible—long before the middle of the twenty-first century—and that electric vehicles need to replace the internal combustion engine in the next ten to fifteen years. Those are big, seemingly impossible benchmarks.

But this is also, honestly, where our next move gets really interesting . . . and potentially revolutionary. The fact of climate change has created the single greatest wealth opportunity in the history of the world. The free-market forge is poised to refine a $90 trillion thing of beauty and wonder by using a "carbon price signal" sufficient to engineer a low-carbon energy economy.

What does it mean to put a price on carbon? Establishing a tax on carbon usage "helps shift the burden for the damage [the external costs of carbon emissions] back to those who are responsible for it, and who can reduce it," says the Carbon Pricing Leadership Coalition. "Instead of dictating who should reduce emissions where and how, a carbon price gives an economic signal and polluters decide for themselves whether to discontinue their polluting activity, reduce emissions, or continue polluting and pay for it."[1] Putting a price on carbon also encourages innovation and clean technology, the organization adds.

The only remaining question is whether American leaders are willing to grab the hammer at the anvil and remake the world.

Entrepreneurs all across the United States are already refining low-carbon energy sources. Utility-scale solar farms are now able to beat other energy power plants on cost, without any government subsidies or incentives.[2] The dependability of battery storage for energy for electric vehicles and national electric grids—the stumbling block that renewable energy critics constantly hold out—is now accelerating at an unprecedented pace.

Even local entrepreneurs are getting in on the act. For instance, a businessman in Alaska may soon produce geothermal energy at a penny per kilowatt hour. That's cheaper than any other energy source, anywhere. The average price of residential electricity in the United States is about twelve cents a kilowatt hour. This Alaskan businessman doesn't care if climate change is real. His low-carbon energy source is all the reality he cares about.[3]

The National Science Foundation launched a national solar research initiative several years ago. It is now leading to critical peer-reviewed science showing how we can store and distribute solar power efficiently and cost effectively in ways that can compete economically with oil-, coal-, and gas-generated power.[4] Would you like to bet against that NSF-led scientific process—the same process that spurred the creation of a company like Google? Not me, especially when the solar industry is *already* beginning to accelerate past a coal industry built on nineteenth-century technology. It's one of the fastest-growing industries in the United States. There are now three times as many solar jobs in America as there are coal mining jobs, though hardly anyone seems to grasp this.[5]

Meanwhile, the wind energy industry is nearly grown up after years of fits and starts. Globally, wind-energy generation has more than quadrupled in the past decade. Energy industry experts objectively predict that, if that pace continues, wind energy sources will power a third of the world's electricity needs within a generation.[6]

A low-carbon economy can be unleashed with a price signal on carbon—which, curiously, is what every major oil, coal, gas, and utility company has been expecting for any number of years now and would compete against once it's set. ExxonMobil, for instance, publicly acknowledged years ago that it was already anticipating a price on carbon and had factored it into its business going forward. Any sort of carbon price signal will do—in any kind of way—that in turn allows the market to respond.[7]

Republican economists, meanwhile, have quietly and clearly refined the ways in which a carbon price signal could generate more than a trillion dollars in new revenue while corporate or personal tax rates are reduced across the board—and almost singlehandedly rescue the United States before it careens off a fiscal cliff. All that truly remains is how to set that price signal on carbon in a way that unleashes the free-market forces of business with an eye on the $90 trillion low-carbon economy prize.

"The idea of using taxes to correct a problem like pollution is an old one with wide support among economists. But it is our unique political moment . . . that may turn the concept into reality," Marty Feldstein and Greg Mankiw (who chaired the White House Council of Economic Advisors under Presidents Ronald Reagan and George W. Bush) wrote in *The New York Times*, explaining why a carbon tax proposal that they and other GOP elder statesmen have now proposed in Washington allows the Republican Party to emerge from the wilderness finally on the climate issue.[8]

We have roughly a decade to get this right, before the inexorable forces of certainty in climate science tell us that we have crossed over into dangerous realms. The usually conservative International Energy Agency (IEA), for instance, has said that the next few years are the drop-dead point of no return for the world to begin to significantly shift the energy mix in the global energy economy. Without that shift in how we obtain energy, the IEA said, we're on our way to a world that's eleven degrees Fahrenheit warmer—which is, understandably, a world of nearly unimaginable horrors for every species trying to survive.[9]

Republican leaders in Congress now have an unprecedented opportunity to let the fiery forge of capitalism work—by placing a carbon price signal on the fiscal table as a trade-off for a cut in corporate and income tax rates and regulatory certainty. GOP economists have shown how a carbon price can remake the economy around energy. Republican-friendly businesses that support the free-market forge and understand the climate threat are already planning for how to compete in a world where a carbon price has been set.

A carbon price of forty dollars a ton that rises over time will change the global energy economy much more rapidly than almost anything else we can imagine—and nearly every industry leader I've ever talked to would support it. "It's really important that we Republicans have a seat at the table when people start talking about climate change," James Baker (who was a secretary of state, a treasury secretary, and White House chief of staff under two Republican presidents) told *The New York Times* in announcing his support for just such a forty-dollars-a-ton carbon price in 2017.[10]

Nearly 150 leading, local newspapers across America—in red and blue states alike—have now endorsed the concept of a price on carbon to begin to significantly change the way in which we produce energy. Many large companies and entrepreneurs are already on board. The only remaining question is: Are America's leaders willing to remake the energy world? To put aside political differences in light of the critical reality and ticking clock we face?

If America acts, the rest of the world will have no choice but to respond. It will, without a doubt, lead to the greatest economic resurgence in history.

A global carbon market is inevitable. The question is whether leaders will allow the historic, fiery anvil of innovation to forge that future before we cross critical tipping points.

PART 6

The Future

We currently face our most important test as a species. To succeed, here's what we must do.

In the course of our long, complex human story, we've had to pass through significant gates to keep moving forward—including surviving the Ice Age, migrating out of Africa, creating civilizations, building machines, exploring and conquering the earth, and enduring two world wars. In spite of all those challenges, we, the human species, survived and made our way through each and every gate.

"The Future" reveals why our new gate—one we created ourselves by exhausting the resources of the earth we've conquered—is the human species' most important test and outlines what we must do to succeed this time around. We *can* make our way through this new gate, too, but how we decide to do it will shape our future as a human species.

As the number of refugees increases—whether from civil wars, persecution, or environmental crises—will the growing trend of nationalism cause countries to seal their borders? Or will leaders and lawmakers unite to solve the real challenges and global threats that all of us commonly share as we coexist under a black sky of our own making?

The critical, important steps we can take right now are manageable, if we work together.

28

Political Morality

The people of Bangladesh have nowhere to go. It is likely to get much worse in the future, on several fronts.

Bangladesh is one of the countries most at risk for even modest sea level rise. If seas rise just one meter, up to 20 percent of the country will be permanently covered in water, forcing millions from their homes.[1]

If (or when) that day arrives, however, Bangladeshis will face a nearly impossible problem. The country is surrounded on three sides by India and on the fourth by the ocean to its south. When Bangladeshis want to leave their country by anything other than an airplane or a boat, they must cross into India in the west, north, and east.

India, however, began to build an eight-foot-high, barbed-wire border fence along the entire length of its 2,544-mile-long, winding, porous border with Bangladesh beginning in the 1980s. It isn't a terribly effective border fence—for now; there are currently about fifteen million Bangladeshis living illegally in India. Many of them made their way across the border through the roughly 20 percent of the border where no wall exists at all.[2]

In the spring of 2017, however—as countries like the United States and Britain began to make nationalist decisions to close their borders to immigrants and refugees fleeing war, persecution, or the threat of environmental collapse in places like Yemen, Somalia, or Sudan—India's political leadership announced that they intended to close the entire border with Bangladesh for good.

India's home minister, Rajnath Singh, said in March 2017 that they intended to completely shut its border with Bangladesh (and Pakistan to its north). "We have decided to seal the border between India and Bangladesh as quickly as possible," Singh told graduates for India's Border Security Force. "I know that some obstacles may arise in this work, as some areas are mountainous, some have jungles, and others have rivers."[3]

Why would India seal its border completely with Bangladesh? The public reason is to keep terrorists out (much as President Trump has promised a southern United States border wall to keep Mexicans and terrorists out). But the real reason, analysts say, is to keep a wave of potential Muslim Bangladeshi refugees from flooding through now-porous borders into outlying areas of India if Bangladesh faces either environmental or economic calamity.

"In the long term, the fence is only going to worsen problems facing the region in general and Bangladesh in particular," says Dr. Sudha Ramachandran, a political and security researcher based in Bangalore, India. "Bangladesh is a low-lying country. A fifth of Bangladesh's territory is likely to go under water if sea levels rise by one meter. There is concern over the fate of Bangladesh's population. India surrounds the country on three sides and the fence is boxing its people in. Some of the most vulnerable coastal districts in Bangladesh are Khulna, Satkhira, and Bagerhat, which lie along India's border. Where will the people go when their homes and crops go under water?"[4]

India's decision to close off its border completely to Bangladesh—whether feasible or not—is a perfect reflection of decisions being made now by world leaders everywhere in the face of growing nationalist calls from voters to seal off their borders as civil wars and environmental calamities drive a flood of refugees to seek new homes.

"In 1989, after the fall of the Berlin Wall, there were only 15 border walls around the world. Today, there are 70 of them," Reuters reported in 2017.[5]

More barriers between countries now exist than at any other time in modern history.

There are now two quite distinct worlds confronting us—both of which revolve around two very different sets of political morality plays. One is inherently selfish and nationalistic; it leads to walls and moats to keep others out. The other recognizes that populations and ecosystems are interconnected in ways that don't recognize borders. Political leaders (like India's home minister or U.S. president Trump) can close borders to keep Bangladeshis or Mexicans out. Or leaders (like German chancellor Angela Merkel or French president Emmanuel Macron) can welcome global pacts such as the Paris Climate Accord that recognize a common threat to the human species requiring unprecedented cooperation.

India is clearly struggling with both of these political morality plays. Even as its leaders promise to close its border with Bangladesh completely to keep refugees from flooding in at some point in the future, other parts of its government have tried to foster exactly the opposite image. India has reached out to neighboring countries, including Bangladesh, with grand plans for seamless travel between the countries in order to boost tourism and trade. They have even proposed transnational roads and rails between the two countries—at the same time they plan to seal off the border entirely.[6]

At some point, India will need to make a decision about which future it plans to embrace: an inward-facing nationalism that closes borders, or a globalist frame that embraces international trade and commerce right alongside deeply challenging (and global) environmental and national security threats.

Security experts like Dr. Ramachandran believe there is only one real choice for political leaders in India and elsewhere who are, only now, waking up to the real challenges we are all beginning to face across the planet.

"India cannot afford to turn a blind eye to the problem," she says.

Not only would that approach be inhumane but also, the impact of rising sea level on India could be as devastating as it is predicted to be on Bangladesh. Indeed, some studies have listed India along with Bangladesh as among the countries at 'extreme risk' from climate change.

Rather than distance itself from Bangladesh on the climate change issue, India should cooperate with it. Taking down the fence is an important first step that Delhi must take. But dismantling walls is more difficult than building them. It requires political will and a change in mindsets. Most of all, it requires recognizing that the India-Bangladesh fence has brought little security to the people of these countries. Rather it is a source of insecurity.[7]

India's choice in the very near future—like so many others—raises a deeply troubling question. Are we approaching the end of morals for nation-states as they choose to protect their citizens from the ravages of environmental collapse at the expense of their neighbors? Leaders will face extremely difficult decisions. Do they close borders and let neighbors starve or foment revolution? As food insecurity takes deeper root, will only those who can afford to pay for their food eat and survive? Is fresh water a right that citizens will demand of their leaders, and will countries like China or India make decisions to protect that right at the expense of neighboring countries?

Global and multinational corporations already confront precisely these sorts of questions routinely now. As we saw with Nestlé, they are preparing for environmental unrest in nearly every country on Earth as water scarcity and fights over natural resources become more omnipresent and public. But when multinational business interests collide with the interests of nation-states, who wins that fight?

I once sat privately with a billionaire at the Four Seasons restaurant in New York as he regaled me with stories of wheeling and dealing over water rights in China and elsewhere. But should a global business profit off fresh water scarcity? Or do the sovereign rights of a nation supersede the rights of that company to profit from that scarcity?

There are wars and rumors of wars in nearly every corner of the earth now. More and more of them mix and match environmental and economic insecurities in some fashion. It isn't a stretch to see India at war with Pakistan over water rights in the not-so-distant future. Yemen, Sudan, and Somalia are examples of what happens to a country when either the water runs out or it becomes impossible to grow enough food to feed people.

If Beijing really does see looming water shortages soon, China may not confine its actions inside its borders. It could very well decide that it has the moral right—a sovereign imperative—to do whatever it must to secure the stuff of life for its own people at the expense of its neighbors.

If the monsoon does, in fact, become unstable and fresh water becomes much more difficult to come by, every country in Asia will face difficult and nearly impossible decisions that will severely test political leaders and strain relations among neighbors.

If the Horn of Africa nations continue to fall further and further into despair over sustained drought, it isn't conceivable that those countries will simply give up and sink into permanent, civil strife that simply can't be managed by political or diplomatic solutions.

Even the United States—the wealthiest nation in the history of the world—may decide that closing borders to both people and trade is the only path for itself. I believe quite strongly that this political path is inherently the wrong, immoral choice, but I also recognize that it is a political choice in the face of looming threats that seem both scary and potentially impossible to deal with.

As environmental impacts and natural resource constraints begin to create seemingly impossible choices, nation-states and their political leaders will be tested in ways that they haven't been before in the course of human history.

But none of those tests are self-contained inside their own borders. The threats they face are global in nature. The solutions to those threats are interconnected and global as well. Hurricanes, droughts, water scarcity, and food insecurity don't recognize borders. Whether political leaders recognize this—and choose to find meaningful ways to cooperate and forge a path out of the box canyon we find ourselves in—is another matter.

29

Our Next Gate

There are big, scary stories in this book, but they are necessary stories. We need to understand that our world is in trouble—right now, at this moment in time, not in some blurry, fuzzy, distant future that no one cares about because we'll all be gone by the time the worst, unimaginable horrors might occur. A world that we have conquered and altered is beginning to turn against us. As a result, *we* are in trouble.

These stories are vital and important because they show us in no uncertain terms that we need to end a vague, misguided notion once and for all—that we are somehow supposed to save the planet. That isn't the challenge before us, the gate through which we must pass.

Water riots led to the collapse of Yemen's government. Saudi Arabia has largely run out of water as well and is now using its oil wealth to feed its people. The monsoon may be on the cusp of instability, threatening a billion people in India who rely on it for their water. Pakistan is poised to go to war with India over access to fresh water. Every country in the Horn of Africa is facing severe water scarcity issues and the people there are facing starvation—just as climate models for the subtropics have long predicted.

China now imports two-thirds of the world's soybeans because it, too, is running out of water in the north as the Gobi Desert swallows the land. It may import *all* of the soybeans on Earth within a decade in order to feed its people. The dozen countries along the southern border of the Sahara Desert in Africa now face a colossal failure of epic proportions because tens of millions of trees they propose to plant to hold back the desert are likely to die within months of being planted.

North Korea is now wholly dependent on food imports for its survival. Half of the people in the world, in fact, now survive on imported food because there isn't enough arable land on which to grow crops to feed the people. The oceans are in far worse shape than most people believe. Half of all species are presently experiencing local extinctions.

Even the United States isn't immune to troubles right now. A Category 6 superstorm could do lasting harm to any coastal city in its path. The groundwater aquifer that runs below California's Central Valley—which is responsible for more than a third of the world's food supply now—is facing many of the same pressures seen in the Arabian, Indus Basin, and Murzuk-Djado Basin aquifers that might be largely depleted now. Extended droughts and extreme precipitation events have the same impact in the American Southwest as they do in the subtropics around the world. Sea level rise along all three coastlines in America may not threaten lives immediately, but they have already rendered many parts of those coastlines uninsurable.

The reasons for the troubles in so many regions of the world are complex, interrelated, and, in nearly all cases, wholly of our own making. Some are tied to overpopulation or land use. Others are examples of human cruelty, stupidity, or warmongering. All are overshadowed by a black sky of our own making, portending a steadily darker future if we don't soon recognize what the harbingers are telling us about Earth's changing system.

But, as we've seen, there are critical, manageable, important steps we can take right now that will give us the best chance to adapt to the changes we're seeing . . . and mitigate damage and suffering for both human beings and other species that share the planet with us.

It starts with common sense.

How is it possible, for instance, that we *still* don't truly know how much water remains in the thirty-seven largest aquifers around the world? Engineers have spent decades determining how much oil is still left underground in places like Saudi Arabia, Iran, Iraq, and Russia. The wealth of major oil and gas multinational corporations like ExxonMobil, Shell, Chevron, BP, and others is built on quantifying and valuing these proven reserves of oil. But somehow, we have no real idea how much water is in the aquifers that literally keep us alive?

The groundwater contained in these aquifers is our second-most important source of fresh water on the planet. They contain roughly a third of the fresh water we need to live. The rest is found at the surface in streams, lakes, rivers, and wetlands or locked away in glaciers and ice caps.[1]

Very recent studies by NASA scientists show that about a third of Earth's largest groundwater basins, especially those that provide fresh water to vast numbers of people in some parts of the world, are being rapidly depleted. What this means is simple: We're using up our water, without knowing when it might run out.

We need to remedy this, right now. If governments won't take on this critical work, then either philanthropies or corporations must. However it's accomplished, the issue must be addressed swiftly. We need to know how much water we have left. We can't simply guess. Current estimates of some of the most critical aquifers are ridiculously and hopelessly useless. In the overstressed Northwest Sahara Aquifer System, for example, what little work has been done to assess when it might run out of water ranges from twenty-one thousand years . . . to ten years.

Jay Famiglietti, NASA's leading expert on water resources, is right. "Available physical and chemical measurements are simply insufficient," he says. "Given how quickly we are consuming the world's groundwater reserves, we need a coordinated global effort to determine how much is left."[2]

As for the fresh water contained in the glaciers that serve as major sources of rivers cascading down the mountains in South Asia, it's long past time that scientists from China and the West begin to talk openly about how to quantify what is truly happening to hundreds of glaciers in the Third Pole. Here, too, we should not have to guess. We need to know what's happening, or else we could see precipitous action to dam up major sources of water in rivers fed by those glaciers that are based on incomplete or even inaccurate information. Pakistan, especially, would benefit from exact knowledge—before either India or China make political decisions that would almost certainly create armed conflict.

Every academic researcher who has set out to study glacier melting in the Himalayas has run into the same brick wall, over and over. The data about the extent of glacier melting in the Himalayas and the Third Pole is inconsistent or nonexistent. It's a patchwork of local knowledge, specific observation when a research team makes the trek to a region, or publicly available satellite data from NASA or the European Space Agency.

Our collective knowledge about how the thousands of small and large glaciers in the region are changing is riddled with gaps. Public satellite images can only provide so much information. They show changes in a given glacier's surface area—but not the glacier's height and mass, which is what matters. Just as knowledge of the *volume* of sea ice in the Arctic is what truly matters—and not how much thin ice might or might not be covering the sea—the same is true of the glaciers across the Himalayas and the Third Pole region. We need to know how much is left there—and up-to-date classified satellite data is more fully capable of providing that information.

While we're at it, we need answers to nearly the exact same question in the Arctic. We don't need to know about ice cover. We need to know how much actual ice is there. We don't know that at present, which is hard to imagine, given the strategic importance of the region to Russia, Canada, and the United States.

The CIA and NSA could largely solve the glacier knowledge gap. Creative glacier researchers have been able to use declassified satellite images over a forty-year time period in order to manually compare them to field data. The ability to compare sparse field data to declassified satellite pictures has helped. It provides the first consistent look, over time, of how the glaciers across the Himalayas are changing—and what this means for tens of millions of people who rely on the glaciers' fresh water as they melt in the spring each year. But the intelligence community has vast image resources at its disposal . . . many more than it makes publicly available to scientists.

Presently they're only used in the service of war and armed conflict. What would matter a great deal is if scientists could see what the generals see.

Sadly, as we've discovered, the CIA shut down its climate center under withering pressure from Republican leaders in Congress. The Pentagon is likewise being asked to stop monitoring distant threats from conflicts over natural resources—precisely as those conflicts are beginning to boil over in countries like Yemen, Syria, Sudan, and elsewhere. But the intelligence community can make its surveillance available to scientists, who will know what to do with such data, if asked.

Right now, scientists are forced to put stakes in the ground at specific locations to measure height and mass of glaciers year in and year out. But there are obviously many, many glaciers in remote parts of the world where that sort of direct observation simply isn't possible. They have no field data in lots and lots of places, which means we are flying blind here, too—just as we are in our knowledge of aquifers. The intelligence community can solve this problem, provided there is the political will to do so.[3]

Why has the United Nations failed to recognize the clear and present dangers now facing environmental refugees beyond carefully worded and creative definitions that clearly have no impact whatsoever in places like the high court of New Zealand? It's incomprehensible that the UN has failed to formally recognize the status of environmental refugees. Bureaucratic mumblings and flowery words have no legal or official status.

Efforts to expand the legal definition of refugees within its 1951 Convention Relating to the Status of Refugees have gone nowhere. The UN's outdated and archaic convention on refugees recognizes economic migrants facing withering poverty and refugees fleeing persecution and war but not environmental refugees caught in a death grip between both. It's beyond insane that the UN's leadership hasn't solved this. It needs to do so, now.[4]

During the marathon talks that culminated in the Paris Climate Accord—the first time in twenty-five years of failure that countries were able to finally cross the finish line on an international climate agreement—the United Nations put its Great Green Wall initiative in the Sahel front and center before five thousand journalists covering the talks.

The initiative, first proposed by the African Union in 2007, was one of the great success stories out of the Paris Climate Accord. The World Bank committed to funding it. So did France. More than twenty countries in Africa have signed up. At least $4 billion was pledged to build a ten-mile-wide swath of trees more than three thousand miles long through a dozen countries along the southern frontier of the Sahara Desert. It was designed to be the largest living structure on the planet.

The Sahel initiative has already claimed success: 15 percent of the trees it's proposed have been planted, largely in Senegal. It claims that four million hectares of land have been restored. But, as we've seen, it might very well be the wrong solution for a very real problem. A first effort to plant fifty million trees in Niger failed. Most of the trees died within months. Only when local farmers in Niger, Burkina Faso, and Mali went back to their local, indigenous roots—by planting trees within crops and learning how to hold water at the base of these trees to conserve water—did the trees serve as a hedge against the advancing desert.

The question now is whether all of the vested partners—from the UN and the World Bank to the African Union—can change course. The solution, as we saw in Niger and Burkina Faso, is a return to indigenous land management at the local level, not a grand plan to put hundreds of millions of trees at the southern edge of the Sahara. One plan will succeed. The other will fail. Can the UN and its financial and country allies recognize the path that local farmers and communities have so clearly charted for them? We'll know soon enough.[5]

Local land management solutions are very clearly at the heart of India's salvation as well. Anupam Mishra's life's work, as discussed earlier, was devoted to just such an approach. Mishra, who died in December 2016, had a vision to properly conserve precious water in India that is directly counter to the grand mega–water projects that India, China, and others are so fond of. But Mishra's approach is the right one. His 2009 TED talk has been viewed now nearly a million times.[6] It needs to be viewed one hundred million times.

The Indian people performed feats of near-magical engineering in the deserts centuries ago in order to harvest and store water, Mishra says. They built ancient aqueducts and stepwells, which work as well today as they did hundreds of years ago.

"Welcome to the Golden Desert," Mishra says at the start of his TED talk. "Clouds seldom visit this area. But we find forty different names of clouds in this dialect used here. There are a number of techniques to harvest rain [but] for the desert society this is no program; this is their life. And they harvest rain in many ways."

Mishra spent his life studying local people and communities behind indigenous, time-honored, and successful rainwater harvesting techniques. What he discovered is that people learned how to develop wells, filter ponds, and catchment systems in the harshest landscapes on Earth. We once learned how to find water for drinking and irrigation in these extremely arid places. We desperately need that knowledge now as India and nearly every country in the subtropics begin to face water scarcity challenges that threaten to tear apart the fabric of life as we know it.

"He saw the callous treatment of water all around him, the pollution of rivers by careless city dwellers and the reckless depletion of groundwater aquifers by farmers with electric-powered tubewells," fellow environmentalist Ramachandra Guha said of Mishra after his death. "So, he began documenting the indigenous systems of water harvesting that were rooted in community control and based on a careful understanding of the local landscape."[7]

Mishra's life's work shows us how to build a bridge between modern water management technology and proven, timeworn methods of harvesting water that is India's legacy and heritage. He was right, when so many others are wrong. We must learn from that work. Our lives, in fact, depend on it now that water scarcity is here to stay in so much of our world.

There are other meaningful, important steps that governments, philanthropies, businesses, educators, and consumers can take. We need dozens of water and infrastructure engineers in Jordan, for instance. We need to bring emerging, breakthrough water desalinization technologies to market as quickly and as cost-effectively as possible. Local knowledge about time-honored indigenous farming techniques needs to spread rapidly to countries like Somalia and other countries in Africa on the edge of ecological collapse.

It's quite difficult—perhaps impossible—to hear stories about the death of great heritage sites like the Great Barrier Reef; the near extinction of iconic species like elephants, giraffes, or mountain apes; or the collapse of pollinators like the rusty-patched bumblebee in North America and not grow terribly despondent. We are witnessing great tragedies and mass extinction events. In some instances, we may be too late. All that remains is to chronicle the end.

But in many instances, there is still much that we can do. Now that we truly know what is happening to our great coral reefs, we can take action. While the biggest changes that we know are necessary—the ones to the sky above us that are warming our oceans—will require a collective force of will that the world's leaders have yet to display, we've seen that there are other smaller actions we can take.

As just one example, we now know that coral fish avoid places where others have left, and overeat the algae in places where everyone has congregated. With that knowledge, we can likely intervene and bring balance back to ecosystems in some coral reefs.[8]

Now that we know the existential threats facing our iconic species, we can preserve them. We shouldn't wait until there are only a few hundred African elephants remaining before moving them to protected spaces. We can assure their ability to continue to walk the plains of Africa. That means that we must recognize the threat that climate changes pose to their habitats, along with threats from poaching, hunting, and development.

Now that hundreds of scientists have sounded the alarm about the collapse of pollinators, some obvious course corrections have become clear. Pollinators clearly fail in monoculture environments. We need to stop requiring them to do their necessary work in massive, single-crop agribusiness environments and provide them with the sorts of diverse environments where they clearly were meant to thrive.

It is also natural, and expected, to be terrified and depressed at the prospect of Category 6 superstorms capable of destroying entire coastal cities in the United States and around the world; over the very real threat that countries like North Korea or Pakistan might strike out militarily against neighbors if water scarcity or food insecurity forces their hand; or the reality that tens of millions of people might die of starvation in countries throughout the subtropics as extreme droughts cause suffering at a scale we haven't seen since the creation of the United Nations after the Second World War.

But, even here, there are actions we can take.

Cities like Tampa Bay or Dubai can plan against such grey swan events. Doing so, right now, will mitigate the damage if (or when) such events occur.

Now that we know North Korea has no hope of growing enough food to feed its people because only 20 percent of its land is now arable, China and the United States must embrace environmental diplomacy. It might succeed here where everything else has failed since the Korean War.

Now that we truly know the scope of the humanitarian crisis facing Somalia, Sudan, Yemen, and every other nation in and around the Horn of Africa, it is impossible to look away. It's not good enough to throw up our hands and bemoan the fact that warlords and dictators use food aid as a weapon against their own people. This is an engineering problem that we *can* fix. We *need* to fix it.

Finally, we now know *precisely* how issues like water scarcity, food insecurity, extreme drought, and other natural resource constraints are intricately connected to civil unrest that inevitably leads to armed conflict, war, terrorism, and other military actions. Yemen, Syria, Egypt, Somalia, and Sudan are all clear examples. Political leaders need to stop ignoring the obvious. To forestall civil unrest in a country, recognize the root causes in the first place. If we do, we may not need annual military budgets of $600 billion.

Throughout the course of our story—the human story—there have been gates we needed to pass through in order to keep moving forward. Our forebears migrated out of Africa as the ice ages eased their death grip on continents. We learned to farm so we no longer needed to hunt for our food. We created civilizations built around great learning. We chose to stop worshipping godlike emperors and kings, assuring ourselves that governments served the people. Science ended great myths, enlightening us all. We learned how to build great machines and artifices that now control every aspect of our environment.

We, human beings, have conquered the earth. We are now its masters. We control every aspect of the planet. There are virtually no places we have not seen or explored. We have been to the bottom of the deepest ocean; to the farthest reaches of the north and south; to the highest mountains; to the center of the blinding deserts and eyes of fierce hurricanes. We have tamed, and killed, every wild beast that ever roamed our darkest nightmares. We have endured not one but two world wars that could have closed yet another gate, but we survived. We, the human species, made our way past each and every challenge, through each and every gate.

We are now standing before another gate, the most difficult one yet in the story of the human species. We have conquered the world. We control every aspect of it. But, in doing so, we have also nearly exhausted its natural resources. We are running out of water and places to grow food for all of us. We have altered our environment in profound ways that we are only beginning to understand. We have darkened the sky above us and changed the oceans between us. And this world, our home, is beginning to turn on us, its master.

This is the gate we must pass through if we are to move forward yet again. We must learn how to live together in the place that we have conquered. If we do not, those things that have defined the human species will begin to end. There are no real physical frontiers left on our planet. We have discovered them all and made those places our own.

The gate that we face at present is not a question of surviving or conquering the planet. We have done all that and more. Now, today, we do not need to save the planet. We have conquered Earth. It is ours. The question before us, the gate we must pass through, is whether we can save ourselves as a vanquished Earth begins to turn against us.

In the end, the planet will be fine. We might not be.

ACKNOWLEDGMENTS

It's always difficult to thank those who've helped with a book's journey. This book, especially, has countless sources of inspiration from the hundreds of scientists, leaders, and colleagues I've had the privilege to learn from over the years. The science—and the obvious conclusions from that research—infuses nearly every paragraph.

But I would like to offer a special thanks to my oldest son, Josh, for helping me understand the concept of humanity's critical next gate and for reading behind me on the extraordinary journey of researching and writing each and every chapter in this book. Thank you to my hardworking agent, Victoria Skurnick, who believes. Thanks as well to the brilliant Tom Dunne and Stephen Power at St. Martin's Press for their thoughtful, strategic guidance that made this book so much better. And, as always, thank you to Ramona Tucker, who is quite possibly the finest editor on Earth.

My professional career is built around a simple concept: facts, science, and evidence matter. Leaders ignore them at their own peril. Democracies fail when truth is distorted or subverted. This book embraces that edifice. We know the truth now. The earth we have vanquished is now turning on its master. The question is what we do from here.

NOTES

Part 1: The Truth
Chapter 1: Einstein's Warning

1. "Einstein's Letter to President Roosevelt—1939," Atomic Archive, http://www.atomicarchive.com/Docs/Begin/Einstein.shtml.

2. Richard G. Hewlett and Oscar E. Anderson, *The New World, 1939–1946* (University Park: Pennsylvania State University Press, 1962): 20–21.

3. "Albert Einstein's Letters to President Franklin Delano Roosevelt," E-World, February 20, 1997, http://hypertextbook.com/eworld/einstein/.

4. "Frisch-Peierls Memorandum," Wikipedia, https://en.wikipedia.org/wiki/Frisch%E2%80%93Peierls_memorandum.

5. "MAUD Committee," Wikipedia, https://en.wikipedia.org/wiki/MAUD_Committee.

6. Hewlett and Anderson, *The New World*, 75.

7. "The Manhattan Project," American Museum of Natural History, http://www.amnh.org/exhibitions/einstein/peace-and-war/the-manhattan-project.

8. "Albert Einstein's Letters to President Franklin Delano Roosevelt."

9. Charles S. Cameron, "Lung Cancer and Smoking: What We Really Know," *Atlantic*, January 1956, https://www.theatlantic.com/magazine/archive/1956/01/lung-cancer-and-smoking-what-we-really-know/304760/.

10. David Kessler, *A Question of Intent: A Great American Battle With a Deadly Industry*, PublicAffairs, http://www.publicaffairsbooks.com/book/a-question-of-intent/david-kessler/9781586481216.

11. "How Smoking Causes Cancer," Cancer Research UK, August 4, 2016, http://

www.cancerresearchuk.org/about-cancer/causes-of-cancer/smoking-and-cancer
/how-smoking-causes-cancer.

12. Thomas Spencer et al., "Up to Four-Fifths of Wetlands Worldwide Could Be at
Risk from Sea Level Rise," University of Cambridge, 2016, http://www.cam.ac.uk
/research/news/up-to-four-fifths-of-wetlands-worldwide-could-be-at-risk-from-sea
-level-rise.

13. Justin Gillis, "A Scientist, His Work, and a Climate Reckoning," *New York Times*,
December 21, 2010, http://www.nytimes.com/2010/12/22/science/earth/22carbon
.html.

14. Charles Keeling's curve, "An Ominous Rise," in Gillis, "A Scientist, His Work,
and a Climate Reckoning."

15. "Past Levels of Carbon Dioxide," "Past Temperatures" charts from Gillis, "A
Scientist, His Work, and a Climate Reckoning."

16. James Hansen, *Making Things Clearer: Exaggeration, Jumping the Gun, and the Venus
Syndrome* (New York: Columbia University, 2013), http://www.columbia.edu/~jeh1
/mailings/2013/20130415_Exaggerations.pdf.

17. "Earth's CO_2 Home Page," CO_2 Earth, https://www.co2.earth.

Chapter 2: Species on the Move

1. Terry L. Root profile, Union of Concerned Scientists, http://www.ucsusa.org
/science-network/member-profiles/terry-root#.WNPan_nyuM8.

2. Terry L. Root, Jeff Price, Kimberly R. Hall, and Alan Pounds, "Fingerprints of
Global Warming on Wild Animals and Plants," *Nature* 421, no. 6918 (2003): 57–60,
https://www.researchgate.net/publication/10964597_Fingerprints_of_global
_warming_on_wild_animals_and_plants.

3. "To What Degree?," Science 360 Video, https://science360.gov/series/degree
/37c32dbf-21f1-4c3b-89ea-ac962cc6efb6.

4. Robert T. Watson, "Climate Change and Biodiversity" (IPCC Technical Paper
V, Intergovernmental Panel on Climate Change, April 2002), https://www.ipcc.ch/pdf
/technical-papers/climate-changes-biodiversity-en.pdf.

5. John J. Wiens, "Climate-Related Local Extinctions Are Already Widespread
among Plant and Animal Species," *PLOS Biology* 14, no. 12 (2016), http://journals.plos
.org/plosbiology/article?id=10.1371/journal.pbio.2001104.

6. Sean Greene, "Hundreds of Species Are Already Going Locally Extinct Because
of Climate Change, Study Says," *Los Angeles Times*, December 13, 2016, http://www
.latimes.com/science/sciencenow/la-sci-sn-climate-change-extinction-20161213-story
.html.

7. Craig Moritz, James L. Patton, Chris J. Conroy, et al., "Impact of a Century of

Climate Change on Small-Mammal Communities in Yosemite National Park, USA," *Science*, October 2008, http://www.sciencemag.org/content/322/5899/261.

8. Morgan W. Tingley, Michelle S. Koo, Craig Moritz, et al., "The Push and Pull of Climate Change Causes Heterogeneous Shifts in Avian Elevational Ranges," *Global Change Biology* 18 (2012): 3279–3290, http://onlinelibrary.wiley.com/doi/10.1111/j.1365 -2486.2012.02784.x/full.

9. Gerardo Ceballos, Paul R. Ehrlich, Anthony D. Barnosky, et al., "Accelerated Modern Human–Induced Species Losses: Entering the Sixth Mass Extinction," *Science Advances*, June 19, 2015, http://advances.sciencemag.org/content/1/5/e1400253.

10. Species on the Move conference, February 2016, http://www.speciesonthemove .com/.

11. "Climate Change 'Forcing Species to Move,'" *Japan Times*, February 11, 2016, http://www.japantimes.co.jp/news/2016/02/11/world/science-health-world/climate -change-forcing-species-move/#.WNO0A_krKM8.

12. "Species on the Move," Inside Climate News, https://insideclimatenews.org /species/archive.

13. Terry L. Root profile.

Chapter 3: The Third Pole

1. Matthew Carney, "Crisis on High," ABC News (Australia), July 25, 2016, http:// www.abc.net.au/news/2016-07-25/climate-change-the-third-pole-under-threat /7657672.

2. Lauren Morello, "'Third Pole' Melting Down, But May Not Diminish Fresh-water Supplies," *Scientific American*, September 13, 2012, https://www.scientificamerican .com/article/third-pole-melting-down-but-may-not-diminish-fresh-water-supplies/.

3. Jane Qiu, "Investigating Climate Change the Hard Way at Earth's Icy 'Third Pole,'" *Scientific American*, June 2, 2016, https://www.scientificamerican.com/article /investigating-climate-change-the-hard-way-at-earth-s-icy-third-pole/.

4. Michael Reilly, "Himalaya Glaciers Melting Much Faster," NBC News, updated November 24, 2008, http://www.nbcnews.com/id/27894721/#.WO1e-1PyulM.

5. Qiu, "Investigating Climate Change."

Chapter 4: Collapse of the Pollinators

1. "Bumble Bees: Rusty Patched Bumble Bee (*Bombus affinis*)," Xerces Society, http:// xerces.org/rusty-patched-bumble-bee/.

2. "The Rusty-Patched Bumble Bee: The Story of a Declining Pollinator," Xerces Society and Day's Edge Productions, http://xerces.maps.arcgis.com/apps/Cascade /index.html?appid=3545656993df4d19a83ffc7987c37c88.

3. Cally Carswell, "Bumblees Aren't Keeping Up with a Warming Planet," *Science*, July 10, 2015, http://science.sciencemag.org/content/sci/349/6244/126.full.pdf.

4. John Schwartz, "Decline of Pollinators Poses Threat to World Food Supply, Report Says," *New York Times*, February 26, 2016, https://www.nytimes.com/2016/02/27/science/decline-of-species-that-pollinate-poses-a-threat-to-global-food-supply-report-warns.html?_r=0.

5. "Deliverable 3(a): Thematic Assessment of Pollinators, Pollination and Food Production," IPBES, March 5, 2017, http://www.ipbes.net/work-programme/pollination.

6. Seth Borenstein, "UN-Sponsored Scientific Report Warns Too Many Species of Pollinators Are Nearing Extinction," *U.S. News*, February 26, 2016, https://www.usnews.com/news/world/articles/2016-02-26/un-science-report-warns-of-fewer-bees-other-pollinators.

7. Schwartz, "Decline of Pollinators."

8. Borenstein, "UN-Sponsored Scientific Report."

9. *The Assessment Report on Pollinators, Pollination and Food Production* (Bonn, Germany: IPBES, 2016), http://www.ipbes.net/sites/default/files/downloads/pdf/spm_deliverable_3a_pollination_20170222.pdf.

10. Carswell, "Bumblees Aren't Keeping Up."

Chapter 5: The "Evil Twin"

1. "Ocean Acidification," Ocean Portal Team, http://ocean.si.edu/ocean-acidification.

2. "What Is Ocean Acidification?," PMEL Carbon Program, https://www.pmel.noaa.gov/co2/story/What+is+Ocean+Acidification%3F.

3. "Ocean Acidification," Ocean Portal Team.

4. "What Is Ocean Acidification?," PMEL Carbon Program.

5. Tim Stephens, "Submarine Springs Reveal How Coral Reefs Respond to Ocean Acidification," University of California, Santa Cruz Newscenter, June 17, 2013, http://news.ucsc.edu/2013/06/calcifying-corals.html.

6. Justin B. Ries, "Shell-Shocked: How Different Creatures Deal with an Acidifying Ocean," *Earth*, March 10, 2010, https://www.earthmagazine.org/article/shell-shocked-how-different-creatures-deal-acidifying-ocean.

7. Stephanie Paige Ogburn, "Ocean Acidification Weakens Mussels' Grips," *Scientific American*, March 13, 2013, https://www.scientificamerican.com/article/ocean-acidification-weakens-mussels-grip/.

8. Elizabeth Grossman, "Northwest Oyster Die-Offs Show Ocean Acidification Has Arrived," *Yale Environment 360*, November 21, 2011, http://e360.yale.edu/features/northwest_oyster_die-offs_show_ocean_acidification_has_arrived.

9. "What Is Ocean Acidification?," PMEL Carbon Program.

10. "Searching for the Ocean Acidification Signal," Ocean Portal, http://ocean.si .edu/ocean-news/searching-ocean-acidification-signal.

11. Tim Flannery, "They're Taking Over!," *New York Review of Books*, September 26, 2013, http://www.nybooks.com/articles/2013/09/26/jellyfish-theyre-taking-over/.

12. Rebecca Albright, Benjamin Mason, Margaret Miller, and Chris Langdon, "Ocean Acidification Compromises Recruitment Success of the Threatened Caribbean Coral *Acropora palmata*," *PNAS* 107, no. 47 (2010): 20400–20404, http://www.pnas.org /content/107/47/20400.abstract.

13. "Effects of Ocean Acidification on Marine Ecosystems," in *Ocean Acidification: A National Strategy to Meet the Challenges of a Changing Ocean* (Washington, D.C.: National Academies Press, 2010), https://www.nap.edu/read/12904/chapter/6.

14. "What are Phytoplankton?," National Ocean Service, jhttp://oceanservice.noaa .gov/facts/phyto.html.

15. Jennifer Chu, "Ocean Acidification May Cause Dramatic Changes to Phyto-plankton," MIT News, July 20, 2015, http://news.mit.edu/2015/ocean-acidification -phytoplankton-0720.

16. "Effects of Ocean Acidification on Marine Ecosystems," #81, https://www.nap .edu/read/12904/chapter/6#81.

17. "Socioeconomic Concerns," in *Ocean Acidification: A National Strategy to Meet the Challenges of a Changing Ocean* (Washington, D.C.: National Academies Press, 2010), https://www.nap.edu/read/12904/chapter/7.

18. "Rising Ocean Acidity May Deplete Vital Phytoplankton," Seeker, http://www .seeker.com/rising-ocean-acidity-may-deplete-vital-phytoplankton-1764998551.html.

Part 2: The Ecosystems
Chapter 6: Regime Shift

1. Paul Huttner, "The Arctic Is Broken: Mild November North Winds in Minne-sota," *Updraft*, November 8, 2016, http://blogs.mprnews.org/updraft/2016/11/the -arctic-is-broken-mild-north-winds-in-minnesota/.

2. John Vidal, "'Extraordinarily Hot' Arctic Temperatures Alarm Scientists," *Guardian*, November 22, 2016, https://www.theguardian.com/environment/2016/nov /22/extraordinarily-hot-arctic-temperatures-alarm-scientists.

3. Fen Montaigne, "An Unusually Warm Arctic Year: Sign of Future Climate Tur-moil?," *Yale Environment 360*, December 5, 2016, http://e360.yale.edu/features/unusually _warm_arctic_climate_turmoil_jennifer_francis.

4. Maria-José Viñas, "See How Arctic Sea Ice Is Losing Its Bulwark Against Warm-ing Summers," NASA, October 28, 2016, updated November 3, 2016, https://www

.nasa.gov/feature/goddard/2016/arctic-sea-ice-is-losing-its-bulwark-against-warming-summers.

5. Marcus Carson and Garry Peterson, eds., *Arctic Resilience Report 2016* (Stockholm, Sweden: Stockholm Environment Institute and the Stockholm Resilience Center, 2016), https://www.sei-international.org/mediamanager/documents/Publications/Arctic ResilienceReport-2016.pdf.

6. Mark D. Ivey, David G. Robinson, Mark B. Boslough, et al., *Arctic Climate Systems Analysis* (Albuquerque, NM: Sandia National Laboratories, March 2015), https://cfwebprod.sandia.gov/cfdocs/CompResearch/docs/SAND20152358.pdf.

7. "The Arctic Is Integral to U.S. National Security," Council on Foreign Relations, March 22, 2017, http://www.cfr.org/polar-regions/arctic-integral-us-national-security/p38955.

Chapter 7: The Sahel

1. Jim Morrison, "The 'Great Green Wall' Didn't Stop Desertification, but It Evolved Into Something That Might," *Smithsonian*, August 23, 2016, http://www.smithsonianmag.com/science-nature/great-green-wall-stop-desertification-not-so-much-180960171/.

2. Kieron Monks, "Can the Great Green Wall Change Direction?," CNN, updated September 26, 2016, http://www.cnn.com/2016/09/22/africa/great-green-wall-sahara/.

3. "Dead Baby Trees by the Millions as Reforestation Fails," IRIN, April 8, 2008, http://www.irinnews.org/news/2008/04/08/dead-baby-trees-millions-reforestation-fails.

4. Morrison, "The 'Great Green Wall.'"

5. Helen Palmer, "Africa's Great Green Wall Is Making Progress on Two Fronts," PRI, May 2, 2016, https://www.pri.org/stories/2016-05-02/africas-great-green-wall-making-progress-two-fronts.

6. Miina Porkka, Joseph H. A. Guillaume, Stefan Siebert, et al., "The Use of Food Imports to Overcome Local Limits to Growth," *Earth's Future* 5, no. 4 (2017): 393–407, http://onlinelibrary.wiley.com/doi/10.1002/2016EF000477/full.

7. Lauren Lipuma, "One-Fifth of World's Population Depends on Food Imports," *GeoSpace*, April 12, 2017, http://blogs.agu.org/geospace/2017/04/12/one-fifth-worlds-population-depends-food-imports.

8. Monks, "Can the Great Green Wall Change Direction?"

9. Morrison, "The 'Great Green Wall.'"

Chapter 8: Ocean Colonies

1. "What Is the Great Barrier Reef?," National Ocean Service, http://oceanservice.noaa.gov/facts/gbrlargeststructure.html.

2. Terry P. Hughes, James T. Kerry, Mariana Alvarez-Noriega, et al., "Global Warming and Recurrent Mass Bleaching of Corals," *Nature* 543, no. 7645 (2017): 373–377, http://www.nature.com/nature/journal/v543/n7645/full/nature21707.html.

3. "Life and Death After Great Barrier Reef Bleaching," Coral Reef Studies, ARC Center of Excellence, November 2016, https://www.coralcoe.org.au/media-releases /life-and-death-after-great-barrier-reef-bleaching.

4. Hughes, Kerry, Alvarez-Noriega, et al., "Global Warming."

5. Damien Cave and Justin Gillis, "Large Sections of Australia's Great Reef Are Now Dead, Scientists Find," *New York Times*, March 15, 2017, https://www.nytimes.com /2017/03/15/science/great-barrier-reef-coral-climate-change-dieoff.html.

6. "The 3rd Global Coral Bleaching Event—2014/2017," Ocean Agency/XL Catlin Seaview Survey, September 23, 2016, http://www.globalcoralbleaching.org/.

7. "Impacts of Mass Coral Die-Off on Indian Ocean Reefs Revealed," Phys.org, February 21, 2017, https://phys.org/news/2017-02-impacts-mass-coral-die-off-indian .html#jCp.

8. Justin Worland, "A Most Beautiful Death: An Underwater Investigation of Coral Bleaching in the South Pacific," *Time*, http://time.com/coral/.

9. Hughes, Kerry, Alvarez-Noriega, et al., "Global Warming."

10. Cave and Gillis, "Large Sections of Australia's Great Reef."

11. "Global Warming and Coral Reefs," National Wildlife Federation, https://www .nwf.org/Wildlife/Threats-to-Wildlife/Global-Warming/Effects-on-Wildlife-and -Habitat/Coral-Reefs.aspx.

12. Tracy D. Ainsworth, Scott F. Heron, Juan Carlos Ortiz, et al., "Climate Change Disables Coral Bleaching Protection on the Great Barrier Reef," *Science*, April 15, 2016, http://science.sciencemag.org/content/352/6283/338.

13. "Coral Reefs," IUCN, https://www.iucn.org/theme/marine-and-polar/get -involved/coral-reefs.

14. "Impacts of Mass Coral Die-Off."

15. Worland, "A Most Beautiful Death."

16. Steve Gittings, "NOAA Scientists Report Mass Die-Off of Invertebrates at East Flower Garden Bank in Gulf of Mexico," National Marine Sanctuaries, July 2016, http://sanctuaries.noaa.gov/news/jul16/noaa-scientists-report-mass-die-off-of -invertebrates-at-east-flower-garden-bank.html.

17. Elena Becatoros, "Scientists Race to Prevent Wipeout of World's Coral Reefs," AP, March 14, 2017, https://apnews.com/b32b7e7d54a047a88751541c9b71314e /scientists-race-prevent-wipeout-worlds-coral-reefs.

18. "Dead Zones May Threaten Coral Reefs Worldwide," *Smithsonian Insider*, May 21, 2017, http://insider.si.edu/2017/03/dead-zones-may-threaten-coral-reefs-worldwide/.

19. Editorial Board, "The Great Barrier Reef Is Dying," *Washington Post*, March 19, 2017, https://www.washingtonpost.com/opinions/the-great-barrier-reef-is-dying/2017/03/19/a1e1277a-0b37-11e7-93dc-00f9bdd74ed1_story.html?utm_term=.f031bff3d03b.

20. "Coral Reefs," IUCN.

21. Ditulis Oleh Susan (November 30, 2010). "Mengenali Sumberdaya Pesisir dan Laut." *Pelajar Progressif* (in Malay). Google, Inc., retrieved April 22, 2013.

22. "Caribbean Coral Reefs May Disappear Within 20 Years: Report," Biharprabha .Com, July 3, 2014, http://news.biharprabha.com/2014/07/caribbean-coral-reefs-may-disappear-within-20-years-report/.

Part 3: The Impacts
Chapter 9: Dome of Heat

1. Jason Samenow, "Iran City Hits Suffocating Heat Index of 154 Degrees Near World Record," *Washington Post*, July 30, 2015, https://www.washingtonpost.com/news/capital-weather-gang/wp/2015/07/30/iran-city-hits-suffocating-heat-index-of-154-degrees-near-world-record/?utm_term=.1fcfcedec568.

2. "Midwest Report: Heat in the Heartland: Climate Change and Economic Risk in the Midwest," Risky Business, http://riskybusiness.org/report/heat-in-the-heartland-climate-change-and-economic-risk-in-the-midwest/.

3. Nick Wiltgen, "Feels-Like Temp Reaches 164 Degrees in Iran, 159 in Iraq; Days Off Ordered as Mideast Broils in Extreme Heat Wave," Weather Channel, August 5, 2015, https://weather.com/news/news/iraq-iran-heat-middle-east-125-degrees.

4. "Research Links Extreme Summer Heat Events to Global Warming," NASA, August 6, 2012, https://www.nasa.gov/topics/earth/features/warming-links.html.

5. "Record High Temps vs. Record Low Temps," Climate Signals, http://www.climatesignals.org/data/record-high-temps-vs-record-low-temps.

6. Jason Samenow, "Two Middle East Locations Hit 129 degrees, Hottest Ever Eastern Hemisphere, Maybe the World," *Washington Post*, July 22, 2016, https://www.washingtonpost.com/news/capital-weather-gang/wp/2016/07/22/two-middle-east-locations-hit-129-degrees-hottest-ever-in-eastern-hemisphere-maybe-the-world/?utm_term=.73d4981cad83.

7. Peter Hannam, "Temperatures Off the Charts as Australia Turns Deep Purple," *Sydney Morning Herald*, January 8, 2013, http://www.smh.com.au/environment/weather/temperatures-off-the-charts-as-australia-turns-deep-purple-20130108-2ce33.html.

8. Damian Carrington, "Australia Adds New Colour to Temperature Maps as Heat Soars," *Guardian*, January 8, 2013, https://www.theguardian.com/environment

/damian-carrington-blog/2013/jan/08/australia-bush-fires-heatwave-temperature
-scale.

9. Stephanie C. Herring, Andrew Hoell, Martin P. Hoerling, et al., "28. Summary
and Broader Context," in "2016: Explaining Extreme Events of 2015 from a Climate
Perspective," *Bulletin of the American Meteorological Society* 97, no. 12 (2016): S141–S145,
http://www.ametsoc.net/eee/2015/28_2015_summary_table.pdf.

10. Alok Jha, "Boiled Alive," *Guardian*, July 26, 2006, http://www.guardian.co.uk
/environment/2006/jul/26/science.g2.

11. Alok Jha, "Climate Change Increased Likelihood of Russian 2010 Heatwave—
Study," *Guardian*, February 21, 2012, http://www.guardian.co.uk/environment/2012
/feb/21/climate-change-russian-heatwave.

12. David Barriopedro, Erich M. Fischer, Jurg Luterbacher, et al., "The Hot Summer of 2010: Redrawing the Temperature Record Map of Europe," *Science*, March 17,
2011, http://www.sciencemag.org/content/early/2011/03/16/science.1201224.

13. Rae Johnston, "Australia's Horrifying Heatwave: Everything You Need to Know,"
Lifehacker, February 10, 2017, https://www.lifehacker.com.au/2017/02/australias
-horrifying-heatwave-everything-you-need-to-know/#6ZQJBPFOuciZOcjl.99.

14. Jaimee Bruce, "Man's Role in 2015 India, Pakistan Heat Wave Confirmed by
Supercomputer Simulations," *Nature World News*, December 21, 2016, http://www
.natureworldnews.com/articles/34271/20161221/mans-role-2015-india-pakistan
-heat-wave-confirmed-supercomputer-simulations.htm.

15. Robert Kopp, Jonathan Buzan, and Matthew Huber, "The Deadly Combination of Heat and Humidity," *New York Times*, June 6, 2015, https://www.nytimes
.com/2015/06/07/opinion/sunday/the-deadly-combination-of-heat-and-humidity
.html.

16. Alex McKinnon, "Australia's New Normal . . . As City Temperatures Hit 47C
People Shelter from the Deadly Heat," *Guardian*, February 18, 2017, https://www
.theguardian.com/australia-news/2017/feb/19/australia-new-normal-47c-climate
-change.

Chapter 10: Category 6

1. "Best Practices for Talking About Climate Change & Hurricanes," Climate
Nexus, http://climatenexus.org/communications-climate-change/message-guide
-climate-storms/.

2. "What Is the Difference Between a Hurricane, a Cyclone, and a Typhoon?,"
National Ocean Service, http://oceanservice.noaa.gov/facts/cyclone.html.

3. Jeff Masters, "Extreme 'Grey Swan' Hurricanes in Tampa Bay: a Potential Future
Disaster," Weather Underground, July 26, 2016, https://www.wunderground.com/blog

/JeffMasters/extreme-grey-swan-hurricanes-in-tampa-bay-a-potential-future-catast
.html.

4. "List of F5 and EF5 Tornadoes," Wikipedia, https://en.wikipedia.org/wiki/List
_of_F5_and_EF5_tornadoes.

5. Masters, "Extreme 'Grey Swan' Hurricanes."

6. Ning Lin and Kerry Emanuel, "Grey Swan Tropical Cyclones," *Nature Climate Change*, August 31, 2015, ftp://texmex.mit.edu/ftp/ftp/pub/emanuel/PAPERS /nclimate2777.pdf.

7. Jennifer Chu, "'Grey Swan' Cyclones Predicted to Be More Frequent and Intense," MIT News, August 31, 2015, http://news.mit.edu/2015/grey-swan-cyclones -storm-surge-0831.

8. Masters, "Extreme 'Grey Swan' Hurricanes."

9. Chu, "'Grey Swan' Cyclones."

10. Brian Donegan, "Is a Category 6 Hurricane Possible?," Weather Channel, July 28, 2016, https://weather.com/storms/hurricane/news/category-6-hurricane -saffir-simpson-wind-scale.

11. Paul Huttner, "Category 6? Extreme Hurricane Patricia Is Strongest in NOAA History," *Updraft*, October 23, 2015, http://blogs.mprnews.org/updraft/2015/10 /category-6-extreme-hurricane-patricia-is-strongest-in-noaa-history/.

12. Andrew Freedman, "Super Typhoon Haima Becomes Earth's Seventh Category 5 Storm of 2016," *Mashable*, October 18, 2016, http://mashable.com/2016/10/18 /super-typhoon-haima-philip.

Chapter 11: The Displaced

1. "Best Practices for Talking About Climate Change & Hurricanes," Climate Nexus, https://climatenexus.org/communications-climate-change/message-guide -climate-storms/.

2. U.S. National Weather Service Lower Mississippi River Forecast Center, "Quality controlled-rainfall analysis for the 2-day period ending August 13th, 2016, at 7:00AM CDT. This time period corresponds to roughly the period of heaviest rainfall over portions of southern Louisiana," Facebook, August 13, 2016, https://www.facebook .com/NWSLMRFC/posts/1145726955498440.

3. "Louisiana Flood August 2016," Climate Signals, updated April 20, 2017, http:// www.climatesignals.org/headlines/events/gulf-storm-august-2016.

4. "Best Practices," Climate Nexus.

5. "Puerto Rico's Power Grid 'Virtually Gone' After Hurricane Maria," EcoWatch, September 28, 2017, https://www.ecowatch.com/puerto-rico-electricity-2490536184 .html.

6. Karin van der Wiel, Sarah B. Kapnick, Geert Jan van Oldenborgh, et al., "Rapid Attribution of the August 2016 Flood-Inducing Extreme Precipitation in South Louisiana to Climate Change," *Hydrology and Earth System Sciences* 21 (2017): 897–921, http://www.hydrol-earth-syst-sci.net/21/897/2017/hess-21-897-2017.pdf.

7. "Louisiana Flooding, Blue Cut Fire Create 100K+ Climate Refugees in the U.S.," Texas Climate Impact, August 19, 2016, http://txclimate.org/100k-climate-related -internally-displaced-persons-in-the-u-s-virtually-overnight/.

8. "Environmental Refugee," OECD, https://stats.oecd.org/glossary/detail.asp?ID =839.

9. "Discussion Note: Migration and the Environment," Ninety-Fourth Session, International Organization for Migration, November 1, 2007, http://www.iom.int/jahia /webdav/shared/shared/mainsite/about_iom/en/council/94/MC_INF_288.pdf.

10. Amy Lieberman, "Where Will the Climate Refugees Go?," Al Jazeera, December 22, 2015, https://www.aljazeera.com/indepth/features/2015/11/climate-refugees -151125093146088.html.

11. *UNHCR Resettlement Handbook* (Geneva, Switzerland: UNHCR, 2011), http:// www.unhcr.org/46f7c0ee2.pdf.

12. Karin Zeitvogel, "50 Million 'Environmental Refugees' by 2020, Experts Say," Phys.org, February 22, 2011, https://phys.org/news/2011-02-million-environmental -refugees-experts.html.

13. François Gemenne, "Climate-Induced Population Displacements in a 4° C+ World," *Philosophical Transactions of the Royal Society A* 369, no. 1934 (2011): 182–195, http://rsta.royalsocietypublishing.org/content/roypta/369/1934/182.full.pdf ?maxtoshow=&hits=10&RESULTFORMAT=1&andorexacttitle=and&andorexact-titleabs=and&andorexactfulltext=and&searchid=1&FIRSTINDEX=0&resource type=HWCIT.

14. Joanna Zelman, "50 Million Environmental Refugees by 2020, Experts Predict," *HuffPost*, February 22, 2011, updated May 25, 2011, http://www.huffingtonpost .com/2011/02/22/environmental-refugees-50_n_826488.html.

15. *Human Tide: The Real Migration Crisis* (London: Christian Aid, 2007), https:// www.christianaid.org.uk/sites/default/files/2017-08/human-tide-the-real-migration -crisis-may-2007.pdf.

16. Michelle Yonetani, *Displacement Due to Natural Hazard-Induced Disasters: Global Estimates for 2009 and 2010* (Geneva, Switzerland: Internal Displacement Monitoring Centre, 2011), http://www.internal-displacement.org/assets/publications/2011/2011 -global-estimates-2009-2010-global-en.pdf.

17. James Hollifield and Idean Salehyan, "Environmental Refugees," Wilson Center, December 21, 2015, https://www.wilsoncenter.org/article/environmental-refugees.

18. Brian Palmer, "There's No Such Thing as a Climate Change Refugee," NRDC, November 16, 2015, https://www.nrdc.org/onearth/theres-no-such-thing-climate -change-refugee.

19. "New Zealand Deports Climate Change Asylum Seeker to Kiribati," BBC News, September 24, 2015, http://www.bbc.com/news/world-asia-34344513.

20. *Teitiota v. Chief Executive of the Ministry of Business Innovation and Employment*, CIV- 2013-404-3528 [2013] N.Z.H.C. 3125. Court case in the high court of New Zealand Auckland Registry between Ioane Teitiota and The Chief Executive of the Ministry of Business Innovation and Employment, hearing: October 16, 2013, judgment: November 26, 2013, https://forms.justice.govt.nz/search/Documents/pdf/jdo/56 /alfresco/service/api/node/content/workspace/SpacesStore/6f4d600a-373f-4ff8 -8ba1-500fb7cc94b0/6f4d600a-373f-4ff8-8ba1-500fb7cc94b0.pdf.

Chapter 12: Vanishing Icons

1. Jacob Stolworthy, "Planet Earth II Crew Witnesses the Tragic Mass Death of Saiga Antelope," *Independent*, November 30, 2016, http://www.independent.co.uk/arts -entertainment/tv/news/planet-earth-ii-crew-witnessed-the-tragic-mass-death-of -saiga-antelope-david-attenborough-bbc1-a7447981.html.

2. Jon Austin, "Mystery as 140,000 Antelopes Drop Dead: Conspiracists Claim Proof of Blood Moon Apocalypse," *Express*, September 8, 2015, http://www.express.co .uk/news/science/603634/Blood-Moon-end-of-the-world-antelope-deaths.

3. "Mass Deaths of Saiga Antelopes in Kazakhstan Linked to Bacteria," CNN, April 15, 2016, http://www.cnn.com/2016/04/15/world/saiga-antelope-mass-deaths -irpt/.

4. E. J. Milner-Gulland, Eric Morgan, Richard Kock, "Planet Earth II: Why More Than 200,000 Saiga Antelopes Died in Just Days," *Conversation*, December 4, 2016, http://theconversation.com/planet-earth-ii-why-more-than-200-000-saiga-antelopes -died-in-just-days-69859.

5. Sonia Altizer, Richard S. Ostfield, Pieter T. J. Johnson, et al., "Climate Change and Infectious Diseases: From Evidence to a Predictive Framework," *Science*, August 2, 2013, http://science.sciencemag.org/content/341/6145/514.full.

6. Shreya Dasgupta, "Saiga Antelope Numbers Rise After Mass Die-Off," *Guardian*, June 20, 2016, https://www.theguardian.com/environment/2016/jun/20/saiga -antelope-numbers-rise-after-mass-die-off.

7. "Mountain Gorilla," WWF, http://wwf.panda.org/what_we_do/endangered _species/great_apes/gorillas/mountain_gorilla/.

8. "Mountain Gorillas and Climate Change," WWF, http://www.worldwildlife.org /pages/mountain-gorillas-and-climate-change.

9. Clara Moskowitz, "Pandas' Bamboo Food May Be Lost to Climate Change," *Live Science*, November 11, 2012, http://www.livescience.com/24697-giant-panda-climate-change-bamboo-impact.html.

10. David McKenzie and Ingrid Formanek, "'Our Living Dinosaurs': There Are Far Fewer African Elephants Than We Thought, Study Shows," CNN, September 1, 2016, http://www.cnn.com/2016/08/31/africa/great-elephant-census/.

11. Melissa de Kock, Wayuphong Jitvijak, Shaun Martin, et al., "African Elephant," WWF Wildlife and Climate Change Series, 2014, https://c402277.ssl.cf1.rackcdn.com/publications/730/files/original/African_elephant_-_WWF_wildlife_and_climate_change_series.pdf?1435158979.

12. "Asian Elephants and Climate Change," WWF, 2017, http://www.worldwildlife.org/pages/asian-elephants-and-climate-change.

13. Alister Doyle, "Giraffes Suffer 'Silent Extinction' in Africa," *Scientific American*, December 8, 2016, https://www.scientificamerican.com/article/giraffes-suffer-silent-extinction-in-africa/.

14. Alister Doyle, "Giraffes Suffer 'Silent Extinction' in Africa: Red List Report," Reuters, December 8, 2016, http://www.reuters.com/article/us-environment-giraffes-idUSKBN13X009.

15. "Big Cats Initiative," *National Geographic*, http://nationalgeographic.org/projects/big-cats-initiative/.

16. David Shukman, "Rhino Poaching: Another Year, Another Grim Record," BBC News, March 9, 2016, http://www.bbc.com/news/science-environment-35769413.

17. Sarah Kaplan, "Cheetahs Are Racing Toward Extinction," *Washington Post*, December 28, 2016, https://www.washingtonpost.com/news/speaking-of-science/wp/2016/12/28/cheetahs-are-racing-toward-extinction/?utm_term=.da6ddb2786be.

Part 4: The Geopolitics
Chapter 13: Seeing Around Corners

1. "DoD Directive 4715.21: Climate Change Adaptation and Resilience," Department of Defense, January 14, 2016, http://www.dtic.mil/whs/directives/corres/pdf/471521p.pdf.

2. Charles Mead and Annie Snider, "Why the CIA Is Spying On a Changing Climate," McClatchy D.C. Bureau, January 10, 2011, http://www.mcclatchydc.com/news/nation-world/world/article24607021.html#storylink=cpy.

3. Mead and Snider, "Why the CIA Is Spying."

4. Jeffrey Goldberg, "The Obama Doctrine," *Atlantic*, April 2016, http://www.theatlantic.com/magazine/archive/2016/04/the-obama-doctrine/471525/.

5. Keith Johnson, "Obama Says Climate Change Is a Security Risk. Why Are Republicans Laughing?," *Foreign Policy*, March 21, 2016, http://foreignpolicy.com/2016/03/21/obama-says-climate-change-is-a-security-risk-why-are-republicans-laughing/.

6. Mead and Snider, "Why the CIA Is Spying."

7. "Security and Climate Change: Are We Living in 'the Age of Consequences'?," Chatham House, December 1, 2016, https://www.chathamhouse.org/event/security-and-climate-change-are-we-living-age-consequences#sthash.3MO8UTjk.dpuf.

8. "Military Experts: Climate Change Could Lead to Humanitarian Crisis," Energy & Climate Intelligence Unit, http://eciu.net/press-releases/2016/military-experts-climate-change-could-lead-to-humanitarian-crisis.

9. Internal Displacement Monitoring Centre, "Grid 2016: Global Report on Internal Displacement," http://www.internal-displacement.org/globalreport2016/.

10. Jared Ferrie, "Climate Change and Mass Migration: A Growing Threat to Global Security," IRIN, January 19, 2017, https://www.irinnews.org/analysis/2017/01/19/climate-change-and-mass-migration-growing-threat-global-security.

11. *Bangladesh Climate Change Strategy and Action Plan 2009* (Dhaka, Bangladesh: Ministry of Environment and Forests, 2009), https://www.iucn.org/downloads/bangladesh_climate_change_strategy_and_action_plan_2009.pdf.

12. Ferrie, "Climate Change and Mass Migration."

OTHER LINKS

https://climateandsecurity.files.wordpress.com/2016/09/climate-and-security-advisory-group_briefing-book-for-a-new-administration_2016_11.pdf

https://www.cna.org/cna_files/pdf/MAB_5-8-14.pdf

https://www.dni.gov/files/documents/SASC_Unclassified_2016_ATA_SFR_FINAL.pdf

https://www.nap.edu/read/14682/chapter/2

https://www.dni.gov/index.php/about/organization/national-intelligence-council-nic-publications

http://archive.defense.gov/pubs/150724-congressional-report-on-national-implications-of-climate-change.pdf?source=govdelivery

https://drive.google.com/file/d/0BzK4C4O6XPJKYk1qUEo2aXUxZ2M/view

http://mashable.com/2016/09/21/national-security-climate-change-risk-obama/#jvlsWYFlOqqH.

Chapter 14: The Water Economy

1. "Nestlé," *Fortune*, http://fortune.com/global500/nestle.

2. Claire Rowan, "The World's Top 100 Food & Beverage Companies—2015:

Change Is the New Normal," *Food Engineering*, September 9, 2015, http://www
.foodengineeringmag.com/articles/94498-the-worlds-top-100-food-beverage
-companies—2015-change-is-the-new-normal.

3. "Tour D'horizon with Nestle: Forget the Global Financial Crisis, the World Is
Running Out of Fresh Water," WikiLeaks, March 24, 2009, https://wikileaks.org/plusd
/cables/09BERN129_a.html.

4. Scott Shane and Andrew W. Lehren, "Leaked Cables Offer Raw Look at U.S.
Diplomacy," *New York Times*, November 28, 2010, http://www.nytimes.com/2010/11
/29/world/29cables.html?pagewanted=all.

5. "The Global Food Crisis" in *The Global Social Crisis: Report on the World Social Situation 2011* (New York: United Nations, 2011), http://www.un.org/esa/socdev/rwss
/docs/2011/chapter4.pdf.

6. John Vidal, "UN Warns of Looming Worldwide Food Crisis in 2013," *Guardian*,
October 13, 2012, http://www.theguardian.com/global-development/2012/oct/14/un
-global-food-crisis-warning.

7. "Tour D'horizon with Nestle," WikiLeaks.

8. "Chrysler the Latest Emissions Cheater?," *Years of Living Dangerously*, http://
yearsoflivingdangerously.com/learn/news/hot-news-january-13-2017/.

9. "Climate Impacts on Water Resources," EPA, January 19, 2017, https://
19january2017snapshot.epa.gov/climate-impacts/climate-impacts-water-resources_
.html.

10. Grantham Institute, Imperial College London, "How Will Climate Change
Impact on Fresh Water Security?," *Guardian*, December 21, 2012, https://www
.theguardian.com/environment/2012/nov/30/climate-change-water.

11. "Water Scarcity," WWF, https://www.worldwildlife.org/threats/water-scarcity.

12. "Water Scarcity," Before the Flood, https://www.beforetheflood.com/explore
/the-crisis/water-scarcity-and-climate-change/.

13. "Water Resources Fall as Water Scarcity Rises on National Agenda," WikiLeaks,
November 14, 2009, https://wikileaks.org/plusd/cables/09SANAA2058_a.html.

14. Hammoud Mounassar, "Clashes Shatter Yemen Truce, US Slams Killings,"
Sydney Morning Herald, May 31, 2011, http://www.smh.com.au//breaking-news-world
/clashes-shatter-yemen-truce-us-slams-killings-20110531-1feb4.html.

15. "Water Scarcity in Yemen: The Country's Forgotten Conflict," *Guardian*, April 2,
2015, http://www.theguardian.com/global-development-professionals-network/2015
/apr/02/water-scarcity-yemen-conflict.

16. "Tour D'horizon with Nestle," WikiLeaks.

Chapter 15: Saudi Arabia

1. Craig S. Smith, "Saudis Worry as They Waste Their Scarce Water," *New York Times,* January 26, 2003, http://www.nytimes.com/2003/01/26/world/saudis-worry-as-they-waste-their-scarce-water.html.

2. Elie Elhadj, "Camels Don't Fly, Deserts Don't Bloom: An Assessment of Saudi Arabia's Experiment in Desert Agriculture" (SOAS/KCL Water Research Group/University of London, May 2004), http://www.soas.ac.uk/water/publications/papers/file38391.pdf.

3. Kai Wellbrock, Peter Voss, and Matthias Grottker, "Reconstruction of Mid-Holocene Climate Conditions for North-Western Arabian Oasis Tayma," *International Journal of Water Resources and Arid Environments* 1, no. 3 (2011): 200–209, http://registerpsipw.org/attachments/article/304/IJWRAE_1(3)200-209.pdf.

4. Smith, "Saudis Worry."

5. Fred Pearce, "Saudi Arabia Stakes a Claim on the Nile," *National Geographic,* December 19, 2012, http://news.nationalgeographic.com/news/2012/12/121217-saudi-arabia-water-grabs-ethiopia/.

6. Nathan Halverson, "What California Can Learn from Saudi Arabia's Water Mystery," *Reveal,* April 22, 2015, https://www.revealnews.org/article/what-california-can-learn-from-saudi-arabias-water-mystery/.

7. Lester R. Brown, "Can the United States Feed China?," Earth Policy Institute, March 23, 2011, http://www.earth-policy.org/plan_b_updates/2011/update93.

8. Halverson, "What California Can Learn."

9. "Almarai Acquires Huge Farmland in Arizona," *Arab News,* March 11, 2014, http://www.arabnews.com/news/537336.

10. "Saudis Invest in Foreign Agriculture for Food Security at Home," WikiLeaks, July 30, 2008, https://wikileaks.org/plusd/cables/08RIYADH1174_a.html.

11. James R. Clapper, "Remarks as Delivered by the Honorable James R. Clapper, Director of National Intelligence," AFCEA/INSA National Security and Intelligence Summit, September 18, 2014, https://www.dni.gov/index.php/newsroom/speeches-interviews/speeches-interviews-2014/item/1115-remarks-as-delivered-by-the-honorable-james-r-clapper-director-of-national-intelligence-afcea-insa-national-security-and-intelligence-summit.

12. Nathan Halverson, "We're Running Out of Water, and the World Powers Are Very Worried," *Reveal,* April 11, 2016, https://www.revealnews.org/article/were-running-out-of-water-and-the-worlds-powers-are-very-worried/.

13. Thomas Buschatzke, "My Turn: Don't Freak Out About Saudi Alfalfa," *Arizona Republic,* November 9, 2015, http://www.azcentral.com/story/opinion/op-ed/2015/11/09/saudi-alfalfa-arizona-water/75475074/.

14. Halverson, "We're Running Out of Water."

Chapter 16: Yemen

1. Farouq al-Kamali, "Could Yemen Become the Next Base for Pirates?," *New Arab*, February 16, 2015, https://www.alaraby.co.uk/english/features/2015/2/16/could -yemen-become-the-next-base-for-pirates.

2. "Yemen," OCHA: United Nations Office for the Coordination of Humanitarian Affairs, http://www.unocha.org/romena/about-us/about-ocha-regional/yemen.

3. Robert Burrowes, "Why Most Yemenis Should Despise Ex-President Ali Abdullah Saleh," *Yemen Times*, February 27, 2012, http://www.yementimes.com/en/1550/opinion /488/Why-most-Yemenis-should-despise—ex-president-Ali-Abdullah-Saleh.htm.

4. Noah Browning, "Yemen Ex-President Amassed up to $60 billion, Colluded with Rebels: U.N. Experts," Reuters, February 25, 2015, http://www.reuters.com/article /us-yemen-security-saleh-un-idUSKBN0LT1C520150225.

5. "Yemen's Saleh Declares Alliance with Houthis," Al Jazeera, May 10, 2015, http://www.aljazeera.com/news/2015/05/cloneofcloneofcloneofstrikes-yemen-saada -breach—150510143647004.html.

6. Judith Evans, "Yemen Could Become First Nation to Run Out of Water," *Times* (of London), October 21, 2009, https://www.thetimes.co.uk/article/yemen-could -become-first-nation-to-run-out-of-water-6jvzddjrl0v.

7. "Time Running Out for Solution to Water Crisis," IRIN, August 13, 2012, http:// www.irinnews.org/report/96093/yemen-time-running-out-for-solution-to-water-crisis.

8. Robert Martin, Al Fry, Eva Haden, and Michael Martin, *Water: Facts and Trends* (Geneva, Switzerland, World Business Council for Sustainable Development, 2009), http://www.wbcsd.org/Clusters/Water/Resources/Water-Facts-and-trends.

9. Ali al-Mujahed and Hugh Naylor, "In Yemen's Grinding War, If the Bombs Don't Get You, the Water Shortages Will," *Washington Post*, July 23, 2015, https://www .washingtonpost.com/world/middle_east/in-yemens-grinding-war-if-the-bombs-dont -get-you-the-water-shortages-will/2015/07/22/a0f60118-299e-11e5-960f -22c4ba982ed4_story.html.

10. "Water Resources Fall as Water Scarcity Rises on National Agenda," WikiLeaks, November 14, 2009, https://wikileaks.org/plusd/cables/09SANAA2058_a.html.

11. "Das Sanderson Hears of Promising Economic and Water Reform Proposals That Need Presidential Push," WikiLeaks, September 22, 2009, https://wikileaks.org /plusd/cables/09SANAA1692_a.html.

Chapter 17: Syria

1. "The Global Food Crisis" in *The Global Social Crisis: Report on the World Social Situation 2011* (New York: United Nations, 2011), http://www.un.org/esa/socdev/rwss /docs/2011/chapter4.pdf.

2. John Vidal, "UN Warns of Looming Worldwide Food Crisis in 2013," *Guardian*, October 13, 2012, http://www.theguardian.com/global-development/2012/oct/14/un-global-food-crisis-warning.

3. Thomas L. Friedman, "WikiLeaks, Drought and Syria," *New York Times*, January 21, 2014, https://www.nytimes.com/2014/01/22/opinion/friedman-wikileaks-drought-and-syria.html?_r=0.

4. "2008 UN Drought Appeal for Syria," WikiLeaks, November 26, 2008, https://wikileaks.org/plusd/cables/08DAMASCUS847_a.html.

5. M. G. Donat, T. C. Peterson, and M. Brunet, "Changes in Extreme Temperature and Precipitation in the Arab Region: Long-Term Trends and Variability Related to ENSO and NAO," *International Journal of Climatology* 34, no. 3 (2013): 581–592, http://onlinelibrary.wiley.com/doi/10.1002/joc.3707/full.

6. Friedman, "WikiLeaks, Drought and Syria."

7. Colin P. Kelley, Shahrzad Mohtadi, Mark A. Cane, Richard Seager, and Yochanan Kushnir, "Climate Change in the Fertile Crescent and Implications of the Recent Syrian Drought," *PNAS* 112, no. 11 (2015): 3241–3246, http://www.pnas.org/content/112/11/3241.

8. Henry Fountain, "Researchers Link Syrian Conflict to a Drought Made Worse by Climate Change," *New York Times*, March 2, 2015, https://www.nytimes.com/2015/03/03/science/earth/study-links-syria-conflict-to-drought-caused-by-climate-change.html.

9. Amy Lieberman, "Where Will the Climate Refugees Go?," Al Jazeera, December 22, 2015, http://www.aljazeera.com/indepth/features/2015/11/climate-refugees-151125093146088.html.

Chapter 18: Jordan

1. "Where We Work," UNRWA, https://www.unrwa.org/where-we-work/jordan.

2. *Water for Life: Jordan's Water Strategy 2008–2022* (Amman, Jordan: Ministry for Water and Irrigation, 2009), http://www.mwi.gov.jo/sites/en-us/Documents/Jordan_Water_Strategy_English.pdf.

3. "Jordanian Farmers Face 'Driest Season in Decades,'" IRIN, February 26, 2014, http://www.irinnews.org/feature/2014/02/26/jordanian-farmers-face-%E2%80%9Cdriest-season-decades%E2%80%9D.

4. *Tapped Out: Water Scarcity and Refugee Pressures in Jordan* (Portland, OR: Mercy Corps, 2014), https://www.mercycorps.org/sites/default/files/MercyCorps_TappedOut_JordanWaterReport_March204.pdf.

5. Valerie Yorke, "Politics Matter: Jordan's Path to Water Security Lies Through Political Reforms and Regional Cooperation" (Working Paper No. 2013/19, NCCR

Trade Regulation, April 2013), http://www.nccr-trade.org/fileadmin/user_upload
/nccr-trade.ch/wp5/5.5a/Valerie_Yorke_NCCR_WP_2013_19_v3.pdf.

6. IRIN, "Jordan Hopes Controversial Red Sea Dead Sea Project Will Stem Water
Crisis," *Guardian*, March 20, 2014, https://www.theguardian.com/global-development
/2014/mar/20/jordan-water-red-sea-dead-sea-project.

7. *Tapped Out*.

Chapter 19: Somalia

1. David Axe, "WikiLeaked Cable Confirms U.S.' Secret Somalia OP," *Wired*,
December 2, 2010, https://www.wired.com/2010/12/wikileaked-cable-confirms-u-s
-secret-somalia-op/.

2. "WikiLeaks Reveals U.S. Twisted Ethiopia's Arm to Invade Somalia," Ayyaan-
tuu News, http://www.ayyaantuu.net/wikileaks-reveals-u-s-twisted-ethiopias-arm-to
-invade-somalia/.

3. Abayomi Azikiwe, "Leaked Cables Confirm U.S. Role in Somalia War," *Workers World*, January 4, 2012, http://www.workers.org/2012/world/somalia_war_0112/.

4. Martin Plaut, "Ethiopia and Kenya Help Dismember Somalia," *New Statesman*,
September 3, 2013, http://www.newstatesman.com/africa/2013/09/ethiopia-and
-kenya-help-dismember-somalia.

5. Ali Abunimah, "US Accused Qatar of Funding Somalia's Al Shabab Militia,
WikiLeaks Reveals," *Electronic Intifada*, August 27, 2011, https://electronicintifada.net
/blogs/ali-abunimah/us-accused-qatar-funding-somalias-al-shabab-militia-wikileaks
-reveals.

6. Abdirisak Itaqile, "Leaked Saudi Cables: Somaliland, Somalia and Puntland
Agreed to Form a Single Government in 2012," *Somaliland Monitor*, June 24, 2015, http://
somalilandmonitor.net/leaked-saudi-cables-somaliland-somalia-and-puntland
-agreed-to-form-a-single-government-in-2012/4977/.

7. Stephen O'Brien, "Under-Secretary-General for Humanitarian Affairs and
Emergency Relief Coordinator, Stephen O'Brien statement to the Security Council on
Missions to Yemen, South Sudan, Somalia, and Kenya and an Update on the Oslo
Conference on Nigeria and the Lake Chad Region," United Nations, March 10, 2017,
https://docs.unocha.org/sites/dms/Documents/ERC_USG_Stephen_OBrien
_Statement_to_the_SecCo_on_Missions_to_Yemen_South_Sudan_Somalia_and
_Kenya_and_update_on_Oslo.pdf.

8. Stephen O'Brien (@UNReliefChief), "We are facing largest human. crisis since
creation of @UN. 20M+ face starvation, famine. W/out global efforts, people will starve
to death," Twitter, 12:20 p.m., March 10, 2017, https://twitter.com/UNReliefChief
/status/840296211814973440.

9. Hilary Clarke, Kara Fox, and Richard Allen Greene, "Alarm Bells Ring for Charities as Trump Pledges to Slash Foreign Aid Budget," CNN, March 1, 2017, http://www.cnn.com/2017/02/28/politics/trump-budget-foreign-aid/.

10. Stephen O'Brien (@UNReliefChief), "In #Somalia, more than 1/2 the population—6.2M people—need humanitarian and protection assistance, incl. 2.9M who are at risk of famine," Twitter, 12:14 p.m., March 10, 2017, https://twitter.com/UNReliefChief/status/840294826696413185.

11. O'Brien, "Under-Secretary-General for Humanitarian Affairs."

12. "Somalis Face Vicious Cycle of Poverty and Desertification," Relief Web, June 18, 2001, http://reliefweb.int/report/somalia/somalis-face-vicious-cycle-poverty-and-desertification.

13. O'Brien, "Under-Secretary-General for Humanitarian Affairs."

Chapter 20: Pakistan

1. Steven Solomon, "Drowning Today, Parched Tomorrow," *New York Times*, August 15, 2010, http://www.nytimes.com/2010/08/16/opinion/16solomon.html.

2. "8 Mighty Rivers Run Dry from Overuse," Nature Animals, http://www.thenatureanimals.com/2012/03/8-mighty-rivers-run-dry-from-overuse.html.

3. William Wheeler, "India and Pakistan at Odds Over Shrinking Indus River," *National Geographic*, October 13, 2011, http://news.nationalgeographic.com/news/2011/10/111012-india-pakistan-indus-river-water/.

4. "Climate Change and Water" (Technical Paper VI, IPCC, June 2008), http://www.ipcc.ch/pdf/technical-papers/climate-change-water-en.pdf.

5. Grantham Institute, Imperial College London, "How Will Climate Change Impact on Fresh Water Security?," *Guardian*, December 21, 2012, https://www.theguardian.com/environment/2012/nov/30/climate-change-water.

6. "Impacts of Climate Change on Water Resources: A Global Perspective," IPCC, http://www.ipcc.ch/ipccreports/tar/wg2/index.php?idp=180.

7. Solomon, "Drowning Today."

8. "Obama's State Department: Putting Water Front and Center," Pulitzer Center, March 18, 2010, http://pulitzercenter.org/blog/obamas-state-department-putting-water-front-and-center.

9. Solomon, "Drowning Today."

10. Wheeler, "India and Pakistan."

11. "2008: Pakistan to Seek Compensation for Diversion of Chenab Waters," *DAWN*, July 2, 2011, https://www.dawn.com/news/640992/2008-pakistan-to-seek-compensation-for-diversion-of-chenab-waters.

12. "2008: Water Scarcity Reigniting Anti-India Sentiment in Pakistan," *DAWN*,

July 2, 2011, https://www.dawn.com/news/640991/2008-water-scarcity-reigniting-anti
-india-sentiment-in-pakistan.

13. "Water Issues Could Sweep Away Indo-Pak Peace Process," *DAWN*, June 21,
2011, https://www.dawn.com/news/638350.

14. Michael Kugelman, "Why the India-Pakistan War over Water Is So Dangerous," *Foreign Policy*, September 30, 2016, http://foreignpolicy.com/2016/09/30/why-the
-india-pakistan-war-over-water-is-so-dangerous-indus-waters-treaty/.

15. *Issues in Managing Water Challenges and Policy Instruments: Regional Perspectives and
Case Studies* (Washington, D.C.: International Monetary Fund, 2015) https://www.imf
.org/external/pubs/ft/sdn/2015/sdn1511tn.pdf.

16. Kugelman, "Why the India-Pakistan War."

Chapter 21: India

1. Shivam Vij, "How Monsoon Rains Lift India's Spirit and Economy," BBC News,
June 9, 2016, http://www.bbc.com/news/world-asia-india-36476535.

2. Eric Bellman, "Stop Blaming the Monsoon: Bad Rains Don't Cause India's Inflation," *Wall Street Journal*, June 6, 2014, https://blogs.wsj.com/indiarealtime/2014
/06/06/stop-blaming-the-monsoon-bad-rains-dont-cause-indias-inflation/.

3. Vij, "How Monsoon Rains Lift."

4. Snehal Fernandes, "Marathwada Drought Man-Made, Not Caused by Climate
Change: Study," *Hindustan Times*, updated October 11, 2016, http://www.hindustantimes
.com/mumbai-news/ht-exclusive-bad-management-of-water-resources-caused
-marathwada-drought/story-fk6Cp5bIduJXw7zDQJ06yL.html.

5. Andy Turner, "The Indian Monsoon in a Changing Climate," Royal Meteorological Society, 2012, https://www.rmets.org/weather-and-climate/climate/indian
-monsoon-changing-climate.

6. Anjali Vaidya, "A Warming Earth Could Mean Monsoons Could Change
Course, and India Must Prepare," *Wire*, September 23, 2015, https://thewire.in/11429
/a-warming-earth-could-mean-monsoons-could-change-course-and-india-must
-prepare/.

7. D. R. Kothawale and Jayashree Revadekar, "Interannual Variations of Indian
Summer Monsoon" (1871–2017), Indian Institute of Tropical Meteorology, http://www
.tropmet.res.in/~kolli/MOL/Monsoon/Historical/air.html.

8. K. N. Krishnakumar, G. S. L. H. V. Prasada Rao, and C. S. Gopakumar, "Rainfall Trends in Twentieth Century over Kerala, India," *Atmospheric Environment* 43,
no. 11 (2009): 1940–1944, http://www.sciencedirect.com/science/article/pii
/S135223100900003X.

9. Mathew Koll Roxy, Kapoor Ritika, Pasal Terray, et al., "Drying of Indian

Subcontinent by Rapid Indian Ocean Warming and a Weakening Land-Sea Thermal Gradient," *Nature Communications* 6 (2015): 7423, https://www.nature.com/articles/ncomms8423.

10. Grace Boyle, "India's Monsoons: A Change in the Rain," Al Jazeera, February 2, 2016, http://www.aljazeera.com/indepth/features/2016/01/india-monsoons-change-rain-160124090758074.html.

11. Roxy, Ritika, Terray, et al., "Drying of Indian Subcontinent."

12. Raghu Murtugudde, "Dust Clouds the Future of the South Asian Monsoon (Op-Ed)," *Live Science*, July 10, 2015, http://www.livescience.com/51520-dust-is-changing-timing-of-monsoons.html.

13. A. S. Sahana, Subimal Ghosh, Auroop Ganguly, and Raghu Murtugudde, "Shift in Indian Summer Monsoon Onset During 1976–1977," *Environmental Research Letters* 10, no. 5 (2015): 054006, http://iopscience.iop.org/1748-9326/10/5/054006/pdf/1748-9326_10_5_054006.pdf.

14. C. T. Sabeerali, Suryachandra A. Rao, R. S. Ajayamohan, and Raghu Murtugudde, "On the Relationship Between Indian Summer Monsoon Withdrawal and Indo-Pacific SST Anomalies Before and After 1976/1977 Climate Shift," *Climate Dynamics* 39, no. 3–4 (2012): 841–859, http://www.tropmet.res.in/awnew/aw-85.pdf.

15. Murtugudde, "Dust Clouds."

16. Elizabeth Howell, "Earth's Groundwater Basins Are Running Out of Water," *Live Science*, July 8, 2015, http://www.livescience.com/51483-groundwater-basins-running-out-of-water.html.

17. Vaidya, "Warming Earth."

18. "Increasing Trend of Extreme Rain Events Over India in a Warming Environment," *Science* 314, no. 5804 (2006): 1442–1445, December 1, 2006, http://www.sciencemag.org/content/314/5804/1442.short.

19. Uttar Pradesh, "Monsoon, or Later," *Economist*, July 28, 2012, http://www.economist.com/node/21559628.

20. Z. A. Thomas, F. Kwasniok, C. A. Boulton, et al., "Early Warnings and Missed Alarms for Abrupt Monsoon Transitions," *Climate of the Past* 11, no. 12 (2015): 1621–1633, http://www.clim-past.net/11/1621/2015/cp-11-1621-2015.pdf.

21. M. Berkelhammer, A. Sinha, L. Stott, H. Cheng, F. S. R. Pausata, and K. Yoshimura, "An Abrupt Shift in the Indian Monsoon 4000 Years Ago," in *Climates, Landscapes, and Civilizations*, ed. L. Giosan, D. Q. Fuller, K. Nicoll, R. K. Flad, and P. D. Clift (Washington, D.C.: American Geophysical Union, 2012), https://www.researchgate.net/profile/Max_Berkelhammer/publication/258553412_An_abrupt_shift_in_the_Indian_Monsoon_4000_years_ago/links/00463528a2010149db000000.pdf.

22. Shri Anupam Mishra, *Aaj Bhi Khare Hain Talaab* [The ponds are still relevant] (New Delhi, India: Gandhi Peace Foundation, 1993), http://www.indiawaterportal.org /articles/aaj-bhi-khare-hain-talaab-book-anupam-mishra.

23. Pradesh, "Monsoon, or Later."

Chapter 22: China

1. Ben Potter, "Will China's Hunger For U.S. Soybeans Last?," AG Web, http:// www.agweb.com/mobile/article/will-chinas-hunger-for-us-soybeans-last-naa-ben -potter/.

2. Huileng Tan, "Soybean Prices Are Surging Thanks to 'Stunning' Demand from China," CNBC, December 20, 2016, http://www.cnbc.com/2016/12/20/soybean -prices-are-surging-thanks-to-stunning-demand-from-china.html.

3. Debra Beachy, "2017 Outlook: Soybeans Eye China Demand vs. South America Supply," AG Web, November 18, 2016, http://www.agweb.com/article/2017 -outlook-soybeans-eye-china-demand-vs-south-america-supply-naa-debra-beachy/.

4. Carrie Brown-Lima, Melissa Cooney, and David Cleary, *An Overview of the Brazil-China Soybean Trade and Its Strategic Implications for Conservation* (Arlington, VA: Nature Conservancy, n.d.), https://www.nature.org/ourinitiatives/regions/latinamerica /brazil/explore/brazil-china-soybean-trade.pdf.

5. Jim Randle, "Little Change at Virginia's China-Owned Smithfield Foods," VOA, September 21, 2015, http://www.voanews.com/a/china-owned-smithfield-foods -growing-little-change/2972885.html.

6. Benjamin Carlson, "Waterless World: China's Ever-Expanding Desert Wasteland," PRI, December 16, 2013, https://www.pri.org/stories/2013-12-16/waterless -world-china-s-ever-expanding-desert-wasteland.

7. Josh Haner, Edward Wong, Derek Watkins, and Jeremy White, "China's Deserts Taking Over as They Expand into Vast Sea of Sand," *Seattle Times*, updated October 24, 2016, http://www.seattletimes.com/nation-world/chinas-deserts-taking-over -as-they-expand-into-vast-sea-of-sand/.

8. Jeremy Luedi, "China's Growing Deserts a Major Political Risk," Global Risk Insights, February 26, 2016, http://globalriskinsights.com/2016/02/chinas-growing -deserts-a-major-political-risk/.

9. "US Embassy Cables: Dalai Lama Says Prioritise Climate Change over Politics in Tibet," *Guardian*, December 16, 2010, https://www.theguardian.com/world/us -embassy-cables-documents/220120.

10. Rob Schmitz, "A Warning for Parched China: A City Runs Out of Water," *Marketplace*, April 25, 2016, https://www.marketplace.org/2016/04/21/world/warning -parched-china-city-runs-out-water.

11. "All Dried Up," *Economist*, October 10, 2013, http://www.economist.com/news/china/21587813-northern-china-running-out-water-governments-remedies-are-potentially-disastrous-all.

12. Schmitz, "A Warning for Parched China."

13. "All Dried Up," *Economist*.

14. "All Dried Up," *Economist*.

15. Tom Phillips, "'Parched' Chinese City Plans to Pump Water from Russian Lake via 1,000 km Pipeline," *Guardian*, March 7, 2017, https://www.theguardian.com/world/2017/mar/07/parched-chinese-city-plans-to-pump-water-from-russian-lake-via-1000km-pipeline.

Chapter 23: Environmental Diplomacy

1. "Department of Defense (DoD) Releases Fiscal Year 2017 President's Budget Proposal," U.S. Department of Defense, February 9, 2016, https://www.defense.gov/News/News-Releases/News-Release-View/Article/652687/department-of-defense-dod-releases-fiscal-year-2017-presidents-budget-proposal/.

2. Dave Gilson, "Why Trump's Military Budget Boost Doesn't Add Up," *Mother Jones*, February 28, 2017, http://www.motherjones.com/politics/2017/02/trump-pentagon-military-budget-spending-charts.

3. "Famine in North Korea," Asia Society, http://asiasociety.org/famine-north-korea.

4. *DPRK Needs and Priorities*, Relief Web, http://reliefweb.int/sites/reliefweb.int/files/resources/DPRK%20Needs%20and%20Priorities%202017.pdf.

5. "North Korea Hunger: Two in Five Undernourished, Says UN," BBC News, March 22, 2017, http://www.bbc.com/news/world-asia-39349726.

6. Oliver Hotham, "The Ups and Downs of Working in North Korean Aid," NKNews.org, February 26, 2016, https://www.nknews.org/2016/02/the-ups-and-downs-of-working-in-north-korean-aid/.

7. "National Intelligence Council—Who We Are," Office of the Director of National Intelligence, https://www.dni.gov/index.php/about/organization/national-intelligence-council-who-we-are.

8. National Intelligence Council, "Southeast Asia: The Impact of Climate Change to 2030: Geopolitical Implications," Homeland Security Digital Library (Conference report CR 2010-02, January 2010), https://www.hsdl.org/?abstract&did=24131.

9. Jessica Benko, "How a Warming Planet Drives Human Migration," *New York Times Magazine*, April 19, 2017, https://www.nytimes.com/2017/04/19/magazine/how-a-warming-planet-drives-human-migration.html.

10. National Intelligence Council, "Southeast Asia."

11. National Intelligence Council, "China: The Impact of Climate Change to 2030: Geopolitical Implications," Homeland Security Digital Library (Conference report CR 2009-09, June 2009), https://www.dni.gov/files/documents/2009%20Conference%20 Report_China_The%20Impact%20of%20Climate%20Change%20to%202030.pdf.

12. "Assessing the Climate Change, Environmental Degradation and Migration Nexus in South Asia," Environmental Migration Portal, http://www.environmental migration.iom.int/projects/assessing-climate-change-environmental-degradation -and-migration-nexus-south-asia.

13. *Assessing the Climate Change, Environmental Degradation and Migration Nexus in South Asia* (Dhaka, Bangladesh: IOM Development Fund, International Organization for Migration, 2016), https://publications.iom.int/system/files/pdf/environmental_degradation _nexus_in_south_asia.pdf.

14. Kim Eunjee, "Experts: N. Korea Especially Vulnerable to Effects of Climate Change," VOA, November 23, 2015, http://www.voanews.com/a/north-korea-climate -change-impact/3071180.html.

15. Steve Etheridge, "Just Like Dad, Kim Jong-un a Boss at Sports," ESPN, April 18, 2012, http://www.espn.com/blog/playbook/fandom/post/_/id/308/kim-jong-un-a -boss-at-sports-too.

16. "How to Deal with the World's Most Dangerous Regime," *Economist*, April 22, 2017, http://www.economist.com/news/leaders/21721146-donald-trump-grapples-his -trickiest-task-how-deal-worlds-most-dangerous-regime.

17. Benjamin Habib, "North Korea and the Global Fight Against Climate Change," *Diplomat*, October 28, 2016, http://thediplomat.com/2016/10/north-korea-and-the -global-fight-against-climate-change/.

Part 5: The Blueprint
Chapter 24: A Path Forward

1. "Introduction," National Science Foundation, https://www.nsf.gov/news/special _reports/climate/intro_background.jsp.

2. "Fifth Assessment Report (AR5)," IPCC, https://www.ipcc.ch/report/ar5/.

3. Jeff Nesbit, "The Earth Breathes," *U.S. News,* January 15, 2013, https://www .usnews.com/news/blogs/at-the-edge/2013/01/15/the-earth-breathes.

4. *Keep It in the Ground* (Oakland, CA: Sierra Club, 2016), https://www.sierraclub .org/sites/www.sierraclub.org/files/blog/Keep%20It%20in%20the%20Ground%20 -%20January%202016.pdf.

5. "National Report: The Economic Risks of Climate Change in the United States," Risky Business, https://riskybusiness.org/report/national/.

6. "Full Report," National Climate Assessment, http://nca2014.globalchange.gov/report.

7. "Fifth Assessment Report (AR5)," IPCC.

8. "The Sustainable Infrastructure Imperative," New Climate Economy, http://newclimateeconomy.report/2016/.

9. "The World Factbook," Central Intelligence Agency, https://www.cia.gov/library/publications/the-world-factbook/.

10. Cory Smith and Laurence Chandy, "Report: How Poor Are America's Poorest? U.S. $2 a Day Poverty in a Global Context," Brookings, August 26, 2014, https://www.brookings.edu/research/how-poor-are-americas-poorest-u-s-2-a-day-poverty-in-a-global-context/.

Chapter 25: Disruption

1. Russ Mitchell and Jessica Meyers, "China Is Banning Traditional Auto Engines. Its Aim: Electric Car Domination," *Los Angeles Times*, September 12, 2017, http://www.latimes.com/business/autos/la-fi-hy-china-vehicles-20170911-story.html.

2. "Fifth Assessment Report (AR5)," IPCC, https://www.ipcc.ch/report/ar5/.

3. "Analysis: Only Five Years Left Before 1.5c Carbon Budget Is Blown," Carbon Brief, May 19, 2016, https://www.carbonbrief.org/analysis-only-five-years-left-before-one-point-five-c-budget-is-blown.

4. *Annual Energy Outlook 2017 with Projections to 2050* (Washington, D.C.: U.S. Energy Information Administration, 2017), https://www.eia.gov/outlooks/aeo/pdf/0383(2017).pdf.

5. Mark Z. Jacobson and Mark A. Delucchi, "A Plan for a Sustainable Future: How to Get All Energy from Wind, Water and Solar Power by 2030," *Scientific American*, November 2009, https://www.scientificamerican.com/media/pdf/200911_Plan_for_Sustainable_Future.pdf.

6. Jacobson and Delucchi, "Plan for a Sustainable Future."

7. *Annual Energy Outlook 2017.*

8. "How Solar Energy Could Be the Largest Source of Electricity by Mid-Century," IEA, September 29, 2014, https://www.iea.org/newsroom/news/2014/september/how-solar-energy-could-be-the-largest-source-of-electricity-by-mid-century.html.

9. "Tesla Gigafactory," Tesla, https://www.tesla.com/gigafactory.

10. Tae Kim, "Tesla to Soar Nearly 50% If Everything Goes Right: Morgan Stanley," CNBC, August 14, 2017, https://www.cnbc.com/2017/08/14/tesla-to-soar-nearly-50-percent-if-everything-goes-right-morgan-stanley.html.

11. "How Solar Energy Could Be."

12. *Renewables 2017 Global Status Report* (Paris: REN21, 2017), http://www.ren21
.net/wp-content/uploads/2017/06/17-8399_GSR_2017_Full_Report_0621_opt.pdf.

13. *JPMorgan Chase to Be 100 Percent Reliant on Renewable Energy by 2020; Announces
$200 Billion Clean Energy Financing Commitment* (New York: JPMorgan Chase, 2017),
https://www.jpmorganchase.com/corporate/Corporate-Responsibility/document
/jpmc-cr-sustainability-fact-sheet.pdf.

14. Lucas Mearian, "U.S. Utilities Face up to $48B Revenue Loss from Solar,
Efficiency," *Computerworld*, December 9, 2014, https://www.computerworld.com
/article/2857672/us-utilities-face-up-to-48b-revenue-loss-from-solar-efficiency
.html.

15. "How Solar Energy Could Be."

16. "New Energy Outlook 2017," *Bloomberg New Energy Finance*, https://about.bnef
.com/new-energy-outlook/.

Chapter 26: Waking the Behemoths

1. "History of HIV and AIDS Overview," Avert, https://www.avert.org
/professionals/history-hiv-aids/overview.

2. Leslie Mullen, "Religion & Astronomy: From Galileo to Aliens," *Space*, https://
www.space.com/10663-astronomy-religion-cosmos-intersection-part2.html.

3. "The Manhattan Project," History, http://www.history.com/topics/the
-manhattan-project.

4. Jessica Lyons Hardcastle, "Field to Market Launches Crop Supply Chain Sustainability Program," *Environmental Leader*, July 8, 2014, https://www.environmental
leader.com/2014/07/field-to-market-launches-crop-supply-chain-sustainability
-program/.

5. "Using the Sun to Power PepsiCo Operations," PepsiCo, June 19, 2014,
http://www.pepsico.com/live/story/using-the-sun-to-power-pepsico-operations
061920141471.

6. "Policy on Climate," General Mills, July 2014, https://www.generalmills.com
/en/News/Issues/climate-policy.

7. "Supporting the White House Climate Data Initiative," Gordon and Betty Moore
Foundation, July 30, 2014, https://www.moore.org/article-detail?newsUrlName
=supporting-the-white-house-climate-data-initiative.

8. "Powering the Future of Wind Energy," Mars, http://www.mars.com/global
/sustainable-in-a-generation/healthy-planet/climate-action/wind-farms.

9. "Assessing Farmer Needs," Nestlé, http://www.nestle.com/csv/communities
/farmer-needs.

10. Dan Mitchell, "Why Monsanto Spent $1 Billion on Climate Data," *Modern*

Farmer, October 2, 2013, https://modernfarmer.com/2013/10/monsanto-spent-1-billion-climate-data/.

11. "Announcing the Innovation Challenge: Using Data Science to Create 'Food Resilience,'" *Microsoft Research Blog,* July 24, 2015, https://www.microsoft.com/en-us/research/blog/announcing-the-innovation-challenge-using-data-science-to-create-food-resilience/.

12. "IBM and Citizen-Scientists Poised to Contribute Equivalent of up to $200 Million for Climate & Environmental Research," IBM, http://www-03.ibm.com/press/us/en/pressrelease/52688.wss.

13. Jeff Barr, "New Amazon Climate Research Grants," Amazon, July 29, 2014, https://aws.amazon.com/blogs/aws/amazon-climate-research-grants/.

14. "Reducing Energy Intensity and Emissions," Walmart, http://corporate.walmart.com/2016grr/enhancing-sustainability/reducing-energy-intensity-and-emissions.

Chapter 27: The Anvil

1. "What Is Carbon Pricing?," Carbon Pricing Leadership Coalition, https://www.carbonpricingleadership.org/what/.

2. Vanessa Dezem and Silvia Martinez, "Argentina Auction Attracts Proposals for 6 GW of Solar, Wind," *Renewable Energy World,* September 12, 2016, http://www.renewableenergyworld.com/articles/2016/09/argentina-auction-attracts-proposals-for-6-gigawatts-of-solar-wind.html.

3. Azadeh Ansan, "'Imagineer' Touts Geothermal Energy Invention," CNN, October 20, 2009, http://www.cnn.com/2009/TECH/science/10/13/geothermal.resort/index.html?iref.

4. "CHE-DMR-DMS Solar Energy Initiative (SOLAR)," National Science Foundation, https://www.nsf.gov/funding/pgm_summ.jsp?pims_id=503298.

5. "Solar Jobs," Solar Energy Industries Association, http://www.seia.org/solar-jobs.

6. "What Is Wind Energy?," Wind Energy Foundation, http://windenergyfoundation.org/what-is-wind-energy/.

7. Jeff Nesbit, "Why ExxonMobil Is Supporting a Carbon Tax Now," *Fortune,* July 10, 2016, http://fortune.com/2016/07/10/exxonmobil-carbon-tax/.

8. Martin S. Feldstein, Ted Halsted, and N. Gregory Mankiw, "A Conservative Case for Climate Action," *New York Times,* February 8, 2017, https://www.nytimes.com/2017/02/08/opinion/a-conservative-case-for-climate-action.html?_r=0.

9. "Climate Change," International Energy Agency, https://www.iea.org/topics/climatechange/.

10. John Schwartz, "A 'Conservative Climate Solution': Republican Group Calls for Carbon Tax," *New York Times*, February 7, 2017, https://www.nytimes.com/2017/02/07/science/a-conservative-climate-solution-republican-group-calls-for-carbon-tax.html.

Part 6: The Future
Chapter 28: Political Morality

1. Sudha Ramachandran,"The India-Bangladesh Wall: Lessons for Trump," *Diplomat*, February 15, 2017, http://thediplomat.com/2017/02/the-india-bangladesh-wall-lessons-for-trump/.

2. Bidisha Banerjee, "The Great Wall of India," *Slate*, December 20, 2010, http://www.slate.com/articles/health_and_science/green_room/2010/12/the_great_wall_of_india.html.

3. Huizhong Wu, "India Wants to Seal Its Borders with Pakistan and Bangladesh," CNN, March 29, 2017, http://www.cnn.com/2017/03/28/asia/india-pakistan-bangladesh-borders/index.html.

4. Ruchi Dua, "As Trump Moves Ahead with the Wall, What Other Countries Have Border Walls? 8 Things to Know," *India Today*, January 27, 2017, http://indiatoday.intoday.in/story/donald-trump-united-states-mexico-border-wall-8-things-to-know/1/867641.html.

5. Dua, "As Trump Moves Ahead."

6. Ramachandran,"The India-Bangladesh Wall."

7. Ramachandran,"The India-Bangladesh Wall."

Chapter 29: Our Next Gate

1. Becky Oskin, "Aquifers: Underground Stores of Freshwater," *Live Science*, January 14, 2015, http://www.livescience.com/39625-aquifers.html.

2. "Study: Third of Big Groundwater Basins in Distress," NASA Jet Propulsion Laboratory, June 16, 2015, https://www.jpl.nasa.gov/news/news.php?feature=4626.

3. Stacy Morford, "Spy Satellites Show the Himalayas' Changing Glaciers," *HuffPost*, December 19, 2016, updated December 20, 2016, http://www.huffingtonpost.com/entry/spy-satellites-show-the-himalayas-changing-glaciers_us_585852c5e4b0630a25423517.

4. James Hollifield and Idean Salehyan, "Environmental Refugees," Wilson Center, December 21, 2015, https://www.wilsoncenter.org/article/environmental-refugees.

5. Kieron Monks, "Can the Great Green Wall Change Direction?," CNN, updated September 26, 2016, http://www.cnn.com/2016/09/22/africa/great-green-wall-sahara/.

6. Anupam Mishra, "The Ancient Ingenuity of Water Harvesting," TED, Decem-

ber 2009, https://www.ted.com/talks/anupam_mishra_the_ancient_ingenuity_of
_water_harvesting/transcript?language=en.

7. Ramachandra Guha, "India: Homage to Anupam Mishra," South Asia Citizens
Web, http://www.sacw.net/article13066.html.

8. Jeff Nesbit, "Why Fish Behavior Should Be Included in Habitat Management,"
Axios, https://www.axios.com/embargoed-until-3pm-coral-reef-fish-choose-to-eat-in
-crowds-2353691494.html.

The bibliography has been posted online at:
https://static.macmillan.com/static/smp/way-the world-ends
/bibliography.html

INDEX